Teacher, Student One-Stop Internet Resources

Log on to bookf.msscience.com

ONLINE STUDY TOOLS
- Section Self-Check Quizzes
- Interactive Tutor
- Chapter Review Tests
- Standardized Test Practice
- Vocabulary PuzzleMaker

ONLINE RESEARCH
- WebQuest Projects
- Prescreened Web Links
- Career Links
- Internet Labs

INTERACTIVE ONLINE STUDENT EDITION
- Complete Interactive Student Edition available at mhln.com

FOR TEACHERS
- Teacher Bulletin Board
- Teaching Today—Professional Development

SAFETY SYMBOLS

SAFETY SYMBOLS	HAZARD	EXAMPLES	PRECAUTION	REMEDY
DISPOSAL	Special disposal procedures need to be followed.	certain chemicals, living organisms	Do not dispose of these materials in the sink or trash can.	Dispose of wastes as directed by your teacher.
BIOLOGICAL	Organisms or other biological materials that might be harmful to humans	bacteria, fungi, blood, unpreserved tissues, plant materials	Avoid skin contact with these materials. Wear mask or gloves.	Notify your teacher if you suspect contact with material. Wash hands thoroughly.
EXTREME TEMPERATURE	Objects that can burn skin by being too cold or too hot	boiling liquids, hot plates, dry ice, liquid nitrogen	Use proper protection when handling.	Go to your teacher for first aid.
SHARP OBJECT	Use of tools or glassware that can easily puncture or slice skin	razor blades, pins, scalpels, pointed tools, dissecting probes, broken glass	Practice common-sense behavior and follow guidelines for use of the tool.	Go to your teacher for first aid.
FUME	Possible danger to respiratory tract from fumes	ammonia, acetone, nail polish remover, heated sulfur, moth balls	Make sure there is good ventilation. Never smell fumes directly. Wear a mask.	Leave foul area and notify your teacher immediately.
ELECTRICAL	Possible danger from electrical shock or burn	improper grounding, liquid spills, short circuits, exposed wires	Double-check setup with teacher. Check condition of wires and apparatus.	Do not attempt to fix electrical problems. Notify your teacher immediately.
IRRITANT	Substances that can irritate the skin or mucous membranes of the respiratory tract	pollen, moth balls, steel wool, fiberglass, potassium permanganate	Wear dust mask and gloves. Practice extra care when handling these materials.	Go to your teacher for first aid.
CHEMICAL	Chemicals can react with and destroy tissue and other materials	bleaches such as hydrogen peroxide; acids such as sulfuric acid, hydrochloric acid; bases such as ammonia, sodium hydroxide	Wear goggles, gloves, and an apron.	Immediately flush the affected area with water and notify your teacher.
TOXIC	Substance may be poisonous if touched, inhaled, or swallowed.	mercury, many metal compounds, iodine, poinsettia plant parts	Follow your teacher's instructions.	Always wash hands thoroughly after use. Go to your teacher for first aid.
FLAMMABLE	Flammable chemicals may be ignited by open flame, spark, or exposed heat.	alcohol, kerosene, potassium permanganate	Avoid open flames and heat when using flammable chemicals.	Notify your teacher immediately. Use fire safety equipment if applicable.
OPEN FLAME	Open flame in use, may cause fire.	hair, clothing, paper, synthetic materials	Tie back hair and loose clothing. Follow teacher's instruction on lighting and extinguishing flames.	Notify your teacher immediately. Use fire safety equipment if applicable.

 Eye Safety Proper eye protection should be worn at all times by anyone performing or observing science activities.

 Clothing Protection This symbol appears when substances could stain or burn clothing.

 Animal Safety This symbol appears when safety of animals and students must be ensured.

 Handwashing After the lab, wash hands with soap and water before removing goggles.

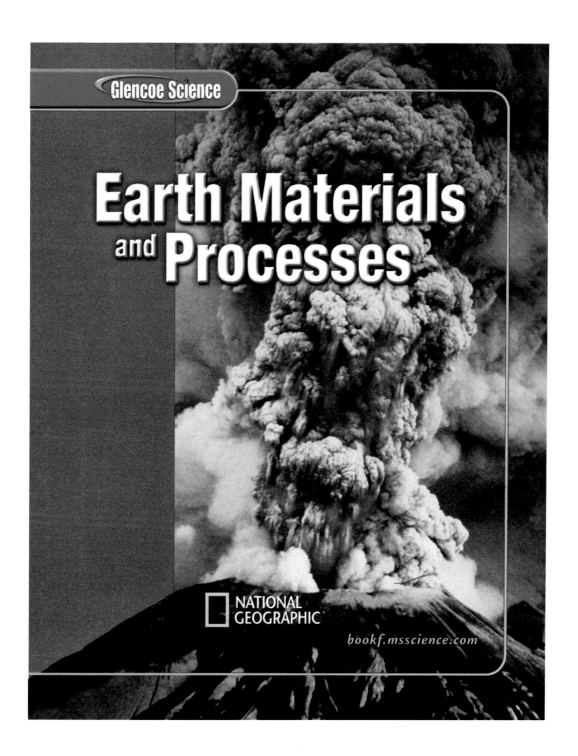

Earth Materials and Processes

New York, New York   Columbus, Ohio   Chicago, Illinois   Peoria, Illinois   Woodland Hills, California

Glencoe Science

Earth Materials and Processes

The eruption column above Mount St. Helens, Washington, as it exploded on May 18, 1980, rose thousands of feet skyward and drifted downwind, dumping dark, gray ash over eastern Washington and beyond. The eruption lasted nine hours, but the landscape was changed within moments.

The McGraw·Hill Companies

Copyright © 2005 by The McGraw-Hill Companies, Inc. All rights reserved. Except as permitted under the United States Copyright Act, no part of this publication may be reproduced or distributed in any form or by any means, or stored in a database or retrieval system, without prior written permission of the publisher.

The National Geographic features were designed and developed by the National Geographic Society's Education Division. Copyright © National Geographic Society. The name "National Geographic Society" and the Yellow Border Rectangle are trademarks of the Society, and their use, without prior written permission, is strictly prohibited.

The "Science and Society" and the "Science and History" features that appear in this book were designed and developed by TIME School Publishing, a division of TIME Magazine. TIME and the red border are trademarks of Time Inc. All rights reserved.

Send all inquiries to:
Glencoe/McGraw-Hill
8787 Orion Place
Columbus, OH 43240-4027

ISBN: 0-07-861749-9

Printed in the United States of America.

4 5 6 7 8 9 10 027/043 09 08 07 06

Authors

Education Division
Washington, D.C.

Ralph M. Feather Jr., PhD
Assistant Professor
Geoscience Department
Indiana University of Pennsylvania
Indiana, PA

Dinah Zike
Educational Consultant
Dinah-Might Activities, Inc.
San Antonio, TX

Series Consultants

CONTENT

William C. Keel, PhD
Department of Physics and Astronomy
University of Alabama
Tuscaloosa, AL

Robert Nierste
Science Department Head
Hendrick Middle School, Plano ISD
Plano, TX

MATH

Michael Hopper, DEng
Manager of Aircraft Certification
L-3 Communications
Greenville, TX

Teri Willard, EdD
Mathematics Curriculum Writer
Belgrade, MT

READING

Carol A. Senf, PhD
School of Literature, Communication, and Culture
Georgia Institute of Technology
Atlanta, GA

SAFETY

Aileen Duc, PhD
Science 8 Teacher
Hendrick Middle School, Plano ISD
Plano, TX

Sandra West, PhD
Department of Biology
Texas State University-San Marcos
San Marcos, TX

ACTIVITY TESTERS

Nerma Coats Henderson
Pickerington Lakeview Jr. High School
Pickerington, OH

Mary Helen Mariscal-Cholka
William D. Slider Middle School
El Paso, TX

Science Kit and Boreal Laboratories
Tonawanda, NY

Series Reviewers

Lois Burdette
Green Bank Elementary-Middle School
Green Bank, WV

Marcia Chackan
Pine Crest School
Boca Raton, FL

Mary Ferneau
Westview Middle School
Goose Creek, SC

Annette D'Urso Garcia
Kearney Middle School
Commerce City, CO

Nerma Coats Henderson
Pickerington Lakeview Jr. High School
Pickerington, OH

Sharon Mitchell
William D. Slider Middle School
El Paso, TX

Joanne Stickney
Monticello Middle School
Monticello, NY

HOW TO...
Use Your Science Book

Why do I need my science book?

Have you ever been in class and not understood all of what was presented? Or, you understood everything in class, but at home, got stuck on how to answer a question? Maybe you just wondered when you were ever going to use this stuff?

These next few pages are designed to help you understand everything your science book can be used for... besides a paperweight!

Before You Read

- **Chapter Opener** Science is occurring all around you, and the opening photo of each chapter will preview the science you will be learning about. The **Chapter Preview** will give you an idea of what you will be learning about, and you can try the **Launch Lab** to help get your brain headed in the right direction. The **Foldables** exercise is a fun way to keep you organized.

- **Section Opener** Chapters are divided into two to four sections. The **As You Read** in the margin of the first page of each section will let you know what is most important in the section. It is divided into four parts. **What You'll Learn** will tell you the major topics you will be covering. **Why It's Important** will remind you why you are studying this in the first place! The **Review Vocabulary** word is a word you already know, either from your science studies or your prior knowledge. The **New Vocabulary** words are words that you need to learn to understand this section. These words will be in **boldfaced** print and highlighted in the section. Make a note to yourself to recognize these words as you are reading the section.

As You Read

- **Headings** Each section has a title in large red letters, and is further divided into blue titles and small red titles at the beginnings of some paragraphs. To help you study, make an outline of the headings and subheadings.

- **Margins** In the margins of your text, you will find many helpful resources. The **Science Online** exercises and **Integrate** activities help you explore the topics you are studying. **MiniLabs** reinforce the science concepts you have learned.

- **Building Skills** You also will find an **Applying Math** or **Applying Science** activity in each chapter. This gives you extra practice using your new knowledge, and helps prepare you for standardized tests.

- **Student Resources** At the end of the book you will find **Student Resources** to help you throughout your studies. These include **Science, Technology,** and **Math Skill Handbooks,** an **English/Spanish Glossary,** and an **Index.** Also, use your **Foldables** as a resource. It will help you organize information, and review before a test.

- **In Class** Remember, you can always ask your teacher to explain anything you don't understand.

FOLDABLES Study Organizer

Science Vocabulary Make the following Foldable to help you understand the vocabulary terms in this chapter.

STEP 1 Fold a vertical sheet of notebook paper from side to side.

STEP 2 Cut along every third line of only the top layer to form tabs.

STEP 3 Label each tab with a vocabulary word from the chapter.

Build Vocabulary As you read the chapter, list the vocabulary words on the tabs. As you learn the definitions, write them under the tab for each vocabulary word.

Look For...

FOLDABLES

At the beginning of every section.

In Lab

Working in the laboratory is one of the best ways to understand the concepts you are studying. Your book will be your guide through your laboratory experiences, and help you begin to think like a scientist. In it, you not only will find the steps necessary to follow the investigations, but you also will find helpful tips to make the most of your time.

- Each lab provides you with a **Real-World Question** to remind you that science is something you use every day, not just in class. This may lead to many more questions about how things happen in your world.

- Remember, experiments do not always produce the result you expect. Scientists have made many discoveries based on investigations with unexpected results. You can try the experiment again to make sure your results were accurate, or perhaps form a new hypothesis to test.

- Keeping a **Science Journal** is how scientists keep accurate records of observations and data. In your journal, you also can write any questions that may arise during your investigation. This is a great method of reminding yourself to find the answers later.

Look For...
- **Launch Labs** start every chapter.
- **MiniLabs** in the margin of each chapter.
- **Two Full-Period Labs** in every chapter.
- **EXTRA Try at Home Labs** at the end of your book.
- the **Web site** with **laboratory demonstrations**.

Before a Test

Admit it! You don't like to take tests! However, there *are* ways to review that make them less painful. Your book will help you be more successful taking tests if you use the resources provided to you.

- Review all of the **New Vocabulary** words and be sure you understand their definitions.
- Review the notes you've taken on your **Foldables,** in class, and in lab. Write down any question that you still need answered.
- Review the **Summaries** and **Self Check questions** at the end of each section.
- Study the concepts presented in the chapter by reading the **Study Guide** and answering the questions in the **Chapter Review.**

Look For...

- Reading Checks and caption questions throughout the text.
- the Summaries and Self Check questions at the end of each section.
- the Study Guide and Review at the end of each chapter.
- the Standardized Test Practice after each chapter.

Let's Get Started

To help you find the information you need quickly, use the Scavenger Hunt below to learn where things are located in Chapter 1.

1. What is the title of this chapter?
2. What will you learn in Section 1?
3. Sometimes you may ask, "Why am I learning this?" State a reason why the concepts from Section 2 are important.
4. What is the main topic presented in Section 2?
5. How many reading checks are in Section 1?
6. What is the Web address where you can find extra information?
7. What is the main heading above the sixth paragraph in Section 2?
8. There is an integration with another subject mentioned in one of the margins of the chapter. What subject is it?
9. List the new vocabulary words presented in Section 2.
10. List the safety symbols presented in the first Lab.
11. Where would you find a Self Check to be sure you understand the section?
12. Suppose you're doing the Self Check and you have a question about concept mapping. Where could you find help?
13. On what pages are the Chapter Study Guide and Chapter Review?
14. Look in the Table of Contents to find out on which page Section 2 of the chapter begins.
15. You complete the Chapter Review to study for your chapter test. Where could you find another quiz for more practice?

Teacher Advisory Board

The Teacher Advisory Board gave the editorial staff and design team feedback on the content and design of the Student Edition. They provided valuable input in the development of the 2005 edition of *Glencoe Science.*

John Gonzales
Challenger Middle School
Tucson, AZ

Rachel Shively
Aptakisic Jr. High School
Buffalo Grove, IL

Roger Pratt
Manistique High School
Manistique, MI

Kirtina Hile
Northmor Jr. High/High School
Galion, OH

Marie Renner
Diley Middle School
Pickerington, OH

Nelson Farrier
Hamlin Middle School
Springfield, OR

Jeff Remington
Palmyra Middle School
Palmyra, PA

Erin Peters
Williamsburg Middle School
Arlington, VA

Rubidel Peoples
Meacham Middle School
Fort Worth, TX

Kristi Ramsey
Navasota Jr. High School
Navasota, TX

Student Advisory Board

The Student Advisory Board gave the editorial staff and design team feedback on the design of the Student Edition. We thank these students for their hard work and creative suggestions in making the 2005 edition of *Glencoe Science* student friendly.

Jack Andrews
Reynoldsburg Jr. High School
Reynoldsburg, OH

Peter Arnold
Hastings Middle School
Upper Arlington, OH

Emily Barbe
Perry Middle School
Worthington, OH

Kirsty Bateman
Hilliard Heritage Middle School
Hilliard, OH

Andre Brown
Spanish Emersion Academy
Columbus, OH

Chris Dundon
Heritage Middle School
Westerville, OH

Ryan Manafee
Monroe Middle School
Columbus, OH

Addison Owen
Davis Middle School
Dublin, OH

Teriana Patrick
Eastmoor Middle School
Columbus, OH

Ashley Ruz
Karrer Middle School
Dublin, OH

The Glencoe middle school science Student Advisory Board taking a timeout at COSI, a science museum in Columbus, Ohio.

Contents

Nature of Science: Monitoring Volcanoes—2

Chapter 1 — Minerals—6

Section 1	Minerals	8
	Lab Crystal Formation	12
Section 2	Mineral Identification	13
Section 3	Uses of Minerals	19
	Lab Mineral Identification	26

Chapter 2 — Rocks—34

Section 1	The Rock Cycle	36
Section 2	Igneous Rocks	40
	Lab Igneous Rock Clues	44
Section 3	Metamorphic Rocks	45
Section 4	Sedimentary Rocks	49
	Lab Sedimentary Rocks	56

Chapter 3 — Earth's Energy and Mineral Resources—64

Section 1	Nonrenewable Energy Resources	66
Section 2	Inexhaustible and Renewable Energy Resources	76
	Lab Soaking up Solar Energy	82
Section 3	Mineral Resources	83
	Lab: Model and Invent Home Sweet Home	88

Chapter 4 — Plate Tectonics—96

Section 1	Continental Drift	98
Section 2	Seafloor Spreading	102
	Lab Seafloor Spreading Rates	105
Section 3	Theory of Plate Tectonics	106
	Lab: Use the Internet Predicting Tectonic Activity	116

In each chapter, look for these opportunities for review and assessment:
- Reading Checks
- Caption Questions
- Section Review
- Chapter Study Guide
- Chapter Review
- Standardized Test Practice
- Online practice at bookf.msscience.com

Contents

Chapter 5: Earthquakes—124

- Section 1 Forces Inside Earth 126
- Section 2 Features of Earthquakes 130
  - Lab Epicenter Location 138
- Section 3 People and Earthquakes 139
  - Lab Earthquake Depths 146

Chapter 6: Volcanoes—154

- Section 1 Volcanoes and Earth's Moving Plates 156
- Section 2 Types of Volcanoes 162
  - Lab Identifying Types of Volcanoes 170
- Section 3 Igneous Rock Features 171
  - Lab: Design Your Own
    How do calderas form? 176

Student Resources

Science Skill Handbook—186
- Scientific Methods 186
- Safety Symbols 195
- Safety in the Science
  Laboratory 196

Extra Try at Home Labs—198

Technology Skill Handbook—201
- Computer Skills 201
- Presentation Skills 204

Math Skill Handbook—205
- Math Review 205
- Science Applications 215

Reference Handbooks—220
- Weather Map Symbols 220
- Rocks 221
- Minerals 222
- Periodic Table of
  the Elements 224

English/Spanish Glossary—227

Index—234

Credits—240

Cross-Curricular Readings/Labs

📹 available as a video lab

NATIONAL GEOGRAPHIC VISUALIZING

1. Crystal Systems 10
2. The Rock Cycle 38
3. Methane Hydrates 72
4. Plate Boundaries 109
5. Visualizing Seismic Waves 132
6. Lava 164

TIME SCIENCE AND Society

2. Australia's Controversial Rock Star 58

TIME SCIENCE AND HISTORY

1. Dr. Dorothy Crowfoot Hodgkin . . . 28

Oops! Accidents in SCIENCE

3. Black Gold! 90
6. Buried in Ash 178

Science and Language Arts

4. Listening In 118

SCIENCE Stats

5. Moving Earth! 148

Launch LAB

1. Distinguish Rocks from Minerals . . 7
2. Observe and Describe Rocks 35
3. Finding Energy Reserves 65
📹 4. Reassemble an Image 97
5. Why do earthquakes occur? 125
6. Map a Volcano 155

Mini LAB

📹 1. Observing Mineral Properties . . . 18
2. Classifying Sediments 50
3. Observing the Effects of Insulation 85
4. Modeling Convection Currents 111
5. Interpreting Seismic Wave Data 135
6. Modeling Volcanic Cones 166

Mini LAB Try at Home

1. Inferring Salt's Crystal System 9
2. Modeling Rock 37
3. Practicing Energy Conservation . . . 73
4. Interpreting Fossil Data 100
📹 5. Modeling Seismic-Safe Structures 144
6. Modeling Magma Movement . . . 160

xii ◆ F

Labs/Activities

One-Page Labs

1 Crystal Formation 13
2 Igneous Rock Clues 44
3 Soaking Up Solar Energy 82
4 Seafloor Spreading Rates 105
5 Epicenter Location 138
6 Identifying Types of Volcanoes . . . 170

Two-Page Labs

1 Mineral Identification 26–27
2 Sedimentary Rocks 56–57
5 Earthquake Depths 146–147

Design Your Own Labs

6 How do calderas form? 176–177

Model and Invent Labs

3 Home Sweet Home 88–89

Use the Internet Labs

4 Predicting Tectonic Activity 116–117

Applying Math

2 Coal Formation 54
5 Earthquake Energy 143
6 Classifying Igneous Rocks 172

Applying Science

1 How can you identify minerals? . . 16
3 Why should you recycle? 86
4 How well do the continents fit together? 108

INTEGRATE

Career: 52, 77, 113, 141, 158
Chemistry: 23, 43, 84, 103
Earth Science: 118, 157
Health: 165
Life Science: 67
Physics: 11, 39, 103, 114, 128, 131
Social Studies: 23

22, 42, 46, 71, 79, 99, 108, 133, 142, 163, 173

Standardized Test Practice

32–33, 62–63, 94–95, 122–123, 152–153, 182–183

F ◆ 1

The Nature of Science

Science Today

Monitoring Volcanoes

Figure 1 The May 18, 1980, eruption of Mount St. Helens blew tons of ash, rock, and steam into the air when it erupted.

Volcanic eruptions can cause incredible destruction, yet many people continue to live near active volcanoes. One approach to protect lives and property is to look for signs that a volcano is about to erupt. This was done on Mount St. Helens in the state of Washington prior to its eruption on May 18, 1980.

Mount St. Helens exploded after 123 years of inactivity. Over 600 km^2 of surrounding land was devastated. More than 300 m of the volcano's north face blew away, creating a huge crater and sending a cloud of hot steam and ash roaring down the flanks of its north slope.

On the island of Hawaii, the Mauna Loa and Kilauea volcanoes erupt more quietly than Mount St. Helens, but they still have the potential to cause great damage. In 1990, lava flows from Kilauea destroyed property in Kalapana Gardens. In 1984, an eruption of Mauna Loa sent lava to within 6.5 km of Hilo, the largest city on the island of Hawaii.

Figure 2 The eruption of Mount St. Helens killed 57 people and caused hundreds of millions of dollars in damage. The force of the blast knocked down millions of trees.

Living Near a Volcano

Volcanoes are natural environmental hazards because of their potentially destructive power and their proximity to populated areas. Many people are reluctant or unwilling to move from their homes near active volcanoes even though there is no way to prevent volcanic eruptions. Such regions often enjoy rich soils of volcanic origin. Consequently, scientists have been working for many years to find the best ways to monitor various volcanoes around the world. They suggest that the data they gather will enable them to better forecast when a quiet volcano might erupt again, allowing people to evacuate a region before an eruption.

Figure 3 Kilauea has erupted continuously for more than 15 years. This lava flow encroached on property in Kalapana Gardens in 1990.

Science

Some advances in the study of volcanoes came about as scientists first attempted to solve the problem of how to forecast eruptions. Solving problems to help make people's lives safer and better is a benefit of science. When you solve a problem by finding a better way to do something, you are doing science.

Volcanology is part of Earth science, the scientific study of the solid part of Earth, the oceans, the atmosphere, and bodies in space. In this book, you will learn about the materials of which Earth is made. You also will learn about processes, such as volcanic eruptions, that shape and change Earth's surface.

Figure 4 Hilo, Hawaii, sits in the path of volcanic lava flows.

The Nature of Science

Science Today

For most of human history, volcanic eruptions have caught people off-guard. Eruptions have poured out lava, hot ash, and gas, often trapping people before they could escape. Today, although eruptions still cause great destruction, fewer people die because volcanologists—scientists who study volcanoes—can forecast many eruptions. For instance, workers knew that Mount St. Helens would explode thanks to advances in volcano monitoring techniques. They were able to warn people in the area and save many lives.

Where Volcanologists Work

Some scientists who monitor volcanoes work at the United States Geological Survey (USGS) volcano observatories, such as:

1. **Alaska Volcano Observatory:** Monitors Alaska's volcanoes and sends out warnings about eruptions in eastern Russia.
2. **Hawaii Volcano Observatory:** Monitors the active volcanoes on the island of Hawaii.
3. **Cascades Volcano Observatory:** Monitors and assesses hazards from volcanoes of the Cascade Range.
4. **Long Valley Observatory:** Monitors activity from the large and potentially hazardous calderas system near Mammoth Lakes, California.

Looking for Signs

Monitoring is reading the signs of activity generated by a volcano before an eruption. For example, prior to a volcanic eruption, magma moves toward Earth's surface. This movement causes earthquakes, changes in a volcano's shape, and the release of certain gases. Volcanologists use specialized instruments to measure changes in the ground surface, the amounts and types of gases emitted, and seismic waves released by earthquakes.

One sign that a volcano might erupt is an increase in the number of earthquakes in the region. Magma and gases force their way up through cracks deep in a volcano, causing the earthquakes. For example, two months before the eruption of Mount St. Helens, about 10,000 quakes occurred in the mountain. Seismographs placed on or near volcanoes can record such earthquakes.

Volcanologists also know that changes in the shape of a volcano can mean an eruption might soon occur. As magma moves upward, parts of a volcano might rise or sink. Mount St. Helens formed a huge bulge in the weeks prior to its eruption.

Using Technology

Besides seismographs, volcanologists use tiltmeters, electronic distance meters (EDMs), spectrometers, and strainmeters. A tiltmeter measures changes in the slope of the ground caused by moving magma. Like a carpenter's level, it consists of a bubble inside a fluid-filled container. If the slope changes, the bubble moves and the difference is measured electronically. An electronic distance meter uses a laser beam to measure the distance between two points on a volcano. If magma moves rocks or widens cracks, the targets will move and the EDM will record a change in distance.

Spectrometers measure gases released from magma. The rate at which volcanoes release carbon dioxide and sulfur dioxide, for example, might change before an eruption.

The strainmeter (or dilatometer) is being used in Hawaii to monitor Mauna Loa and Kilauea. It consists of a small canister filled with liquid silicon that is placed deep in a hole drilled into a volcano. Any movement in the volcano that changes the shape of the ground squeezes the strainmeter and the measurements are recorded on instruments at the surface.

Figure 5 This USGS solar-powered seismograph records small earthquakes on the flank of the Augustine volcano in Alaska.

Working on a Volcano

Although some volcanoes are monitored using radio-controlled instruments, volcanologists also must work in dangerous conditions on active volcanoes. They install instruments, take readings, or collect gas escaping from volcanic vents.

Volcanologist Cynthia Gardner enjoys her work in Washington, Oregon, and Alaska because she's helping to save lives. When she's not in the field, she collects data, writes reports, and sets up emergency procedures in communities near volcanoes.

Figure 6 Volcanologist Cynthia Gardner uses advanced equipment to monitor volcanoes.

Airplanes and satellites are tools that help volcanologists forecast the eruption of volcanoes. Research in your local library or by visiting bookf.msscience.com to find out how volcanologists employ these tools. How would their work be more difficult without the aid of airplanes and satellites?

chapter 1

Minerals

chapter preview

sections

1. **Minerals**
   Lab Crystal Formation
2. **Mineral Identification**
3. **Uses of Minerals**
   Lab Mineral Identification

   Virtual Lab How can minerals be defined by their properties?

Nature's Beautiful Creation

Although cut by gemologists to enhance their beauty, these gorgeous diamonds formed naturally—deep within Earth. One requirement for a substance to be a mineral is that it must occur in nature. Human-made diamonds serve their purpose in industry but are not considered minerals.

Science Journal Write two questions you would ask a gemologist about the minerals that he or she works with.

Start-Up Activities

Distinguish Rocks from Minerals

When examining rocks, you'll notice that many of them are made of more than one material. Some rocks are made of many different crystals of mostly the same mineral. A mineral, however, will appear more like a pure substance and will tend to look the same throughout. Can you tell a rock from a mineral?

1. Use a magnifying lens to observe a quartz crystal, salt grains, and samples of sandstone, granite, calcite, mica, and schist (SHIHST).
2. Draw a sketch of each sample.
3. Infer which samples are made of one type of material and should be classified as minerals.
4. Infer which samples should be classified as rocks.
5. **Think Critically** In your Science Journal, compile a list of descriptions for the minerals you examined and a second list of descriptions for the rocks. Compare and contrast your observations of minerals and rocks.

Minerals Make the following Foldable to help you better understand minerals.

STEP 1 Fold a vertical sheet of notebook paper from side to side.

STEP 2 Cut along every third line of only the top layer to form tabs.

STEP 3 Label each tab with a question.

Ask Questions Before you read the chapter, write questions you have about minerals on the front of the tabs. As you read the chapter, add more questions and write answers under the appropriate tabs.

Preview this chapter's content and activities at bookf.msscience.com

section 1
Minerals

as you read

What You'll Learn
- **Describe** characteristics that all minerals share.
- **Explain** how minerals form.

Why It's Important
You use minerals and products made from them every day.

Review Vocabulary
atoms: tiny particles that make up matter; composed of protons, electrons, and neutrons

New Vocabulary
- mineral
- crystal
- magma
- silicate

Figure 1 You probably use minerals or materials made from minerals every day without thinking about it.
Infer How many objects in this picture might be made from minerals?

What is a mineral?

How important are minerals to you? Very important? You actually own or encounter many things made from minerals every day. Ceramic, metallic, and even some paper items are examples of products that are derived from or include minerals. **Figure 1** shows just a few of these things. Metal bicycle racks, bricks, and the glass in windows would not exist if it weren't for minerals. A **mineral** is a naturally occurring, inorganic solid with a definite chemical composition and an orderly arrangement of atoms. About 4,000 different minerals are found on Earth, but they all share these four characteristics.

Mineral Characteristics First, all minerals are formed by natural processes. These are processes that occur on or inside Earth with no input from humans. For example, salt formed by the natural evaporation of seawater is the mineral halite, but salt formed by evaporation of saltwater solutions in laboratories is not a mineral. Second, minerals are inorganic. This means that they aren't made by life processes. Third, every mineral is an element or compound with a definite chemical composition. For example, halite's composition, NaCl, gives it a distinctive taste that adds flavor to many foods. Fourth, minerals are crystalline solids. All solids have a definite volume and shape. Gases and liquids like air and water have no definite shape, and they aren't crystalline. Only a solid can be a mineral, but not all solids are minerals.

Atom Patterns The word *crystalline* means that atoms are arranged in a pattern that is repeated over and over again. For example, graphite's atoms are arranged in layers. Opal, on the other hand, is not a mineral in the strictest sense because its atoms are not all arranged in a definite, repeating pattern, even though it is a naturally occurring, inorganic solid.

Figure 2 More than 200 years ago, the smooth, flat surfaces on crystals led scientists to infer that minerals had an orderly structure inside.

Even though this rose quartz looks uneven on the outside, its atoms have an orderly arrangement on the inside.

The well-formed crystal shapes exhibited by these clear quartz crystals suggest an orderly structure.

The Structure of Minerals

Do you have a favorite mineral sample or gemstone? If so, perhaps it contains well-formed crystals. A **crystal** is a solid in which the atoms are arranged in orderly, repeating patterns. You can see evidence for this orderly arrangement of atoms when you observe the smooth, flat outside surfaces of crystals. A crystal system is a group of crystals that have similar atomic arrangements and therefore similar external crystal shapes.

Reading Check *What is a crystal?*

Crystals Not all mineral crystals have smooth surfaces and regular shapes like the clear quartz crystals in **Figure 2.** The rose quartz in the smaller photo of **Figure 2** has atoms arranged in repeating patterns, but you can't see the crystal shape on the outside of the mineral. This is because the rose quartz crystals developed in a tight space, while the clear quartz crystals developed freely in an open space. The six-sided, or hexagonal crystal shape of the clear quartz crystals in **Figure 2,** and other forms of quartz can be seen in some samples of the mineral. **Figure 3** illustrates the six major crystal systems, which classify minerals according to their crystal structures. The hexagonal system to which quartz belongs is one example of a crystal system.

Crystals form by many processes. Next, you'll learn about two of these processes—crystals that form from magma and crystals that form from solutions of salts.

Mini LAB

Inferring Salt's Crystal System

Procedure
1. Use a **magnifying lens** to observe grains of common **table salt** on a dark sheet of **construction paper.** Sketch the shape of a salt grain. **WARNING:** *Do not taste or eat mineral samples. Keep hands away from your face.*
2. Compare the shapes of the salt crystals with the shapes of crystals shown in **Figure 3.**

Analysis
1. Which characteristics do all the grains have in common?
2. Research another mineral with the same crystal system as salt. What is this crystal system called?

Try at Home

NATIONAL GEOGRAPHIC VISUALIZING CRYSTAL SYSTEMS

Figure 3

A crystal's shape depends on how its atoms are arranged. Crystal shapes can be organized into groups known as crystal systems—shown here in 3-D with geometric models (in blue). Knowing a mineral's crystal system helps researchers understand its atomic structure and physical properties.

▲ **CUBIC** Fluorite is an example of a mineral that forms cubic crystals. Minerals in the cubic crystal system are equal in size along all three principal dimensions.

▲ **HEXAGONAL** (hek SA guh nul) In hexagonal crystals, horizontal distances between opposite crystal surfaces are equal. These crystal surfaces intersect to form 60° or 120° angles. The vertical length is longer or shorter than the horizontal lengths.

◀ **TETRAGONAL** (te TRA guh nul) Zircon crystals are tetragonal. Tetragonal crystals are much like cubic crystals, except that one of the principal dimensions is longer or shorter than the other two dimensions.

▲ **ORTHORHOMBIC** (awr thuh RAHM bihk) Minerals with orthorhombic structure, such as barite, have dimensions that are unequal in length, resulting in crystals with a brick-like shape.

▲ **MONOCLINIC** (mah nuh KLIH nihk) Minerals in the monoclinic system, such as orthoclase, also exhibit unequal dimensions in their crystal structure. Only one right angle forms where crystal surfaces meet. The other angles are oblique, which means they don't form 90° angles where they intersect.

▲ **TRICLINIC** (tri KLIH nihk) The triclinic crystal system includes minerals exhibiting the least symmetry. Triclinic crystals, such as rhodonite (ROH dun ite), are unequal in all dimensions, and all angles where crystal surfaces meet are oblique.

Figure 4 Minerals form by many natural processes.

A This rock formed as magma cooled slowly, allowing large mineral grains to form.

Labradorite

B Some minerals form when salt water evaporates, such as these white crystals of halite in Death Valley, California.

Crystals from Magma Natural processes form minerals in many ways. For example, hot melted rock material, called **magma,** cools when it reaches Earth's surface, or even if it's trapped below the surface. As magma cools, its atoms lose heat energy, move closer together, and begin to combine into compounds. During this process, atoms of the different compounds arrange themselves into orderly, repeating patterns. The type and amount of elements present in a magma partly determine which minerals will form. Also, the size of the crystals that form depends partly on how rapidly the magma cools.

When magma cools slowly, the crystals that form are generally large enough to see with the unaided eye, as shown in **Figure 4A.** This is because the atoms have enough time to move together and form into larger crystals. When magma cools rapidly, the crystals that form will be small. In such cases, you can't easily see individual mineral crystals.

Crystals from Solution Crystals also can form from minerals dissolved in water. When water evaporates, as in a dry climate, ions that are left behind can come together to form crystals like the halite crystals in **Figure 4B.** Or, if too much of a substance is dissolved in water, ions can come together and crystals of that substance can begin to form in the solution. Minerals can form from a solution in this way without the need for evaporation.

INTEGRATE Physics

Crystal Formation Evaporites commonly form in dry climates. Research the changes that take place when a saline lake or shallow sea evaporates and halite or gypsum forms.

Mineral Compositions and Groups

Ninety elements occur naturally in Earth's crust. Approximately 98 percent (by weight) of the crust is made of only eight of these elements, as shown in **Figure 5.** Of the thousands of known minerals, only a few dozen are common, and these are mostly composed of the eight most common elements in Earth's crust.

Most of the common rock-forming minerals belong to a group called the silicates. **Silicates** (SIH luh kayts) are minerals that contain silicon (Si) and oxygen (O) and usually one or more other elements. Silicon and oxygen are the two most abundant elements in Earth's crust. These two elements alone combine to form the basic building blocks of most of the minerals in Earth's crust and mantle. Feldspar and quartz, which are silicates, and calcite, which is a carbonate, are examples of common, rock-forming minerals. Other mineral groups also are defined according to their compositions.

Figure 5 Most of Earth's crust is composed of eight elements.

section 1 review

Summary

What is a mineral?
- Many products used by humans are made from minerals.
- Minerals are defined by four main characteristics.

The Structure of Minerals
- The crystal shape of a mineral reflects the way in which its atoms are arranged.
- Minerals are classified according to the types of atoms in their structures and the way that the atoms are arranged.

Mineral Compositions and Groups
- Only eight elements form approximately 98 percent (by weight) of Earth's crust.
- The majority of Earth's crust is composed of silicate minerals.

Self Check

1. **List** four characteristics that all minerals share.
2. **Describe** two ways that minerals can form from solution.
3. **Explain** whether diamonds made in the laboratory are considered to be minerals.
4. **Describe** how crystals of minerals are classified.
5. **Think Critically** The mineral dolomite, a rock-forming mineral, contains oxygen, carbon, magnesium, and calcium. Is dolomite a silicate? Explain.

Applying Skills

6. **Graph** Make a graph of your own design that shows the relative percentages of the eight most common elements in Earth's crust. Then determine the approximate percentage of the crust that is made up of iron and aluminum. If one is available, you may use an electronic spreadsheet program to make your graph and perform the calculation.

Crystal Formation

In this lab, you'll have a chance to learn how crystals form from solutions.

● Real-World Question
How do crystals form from solution?

Goals
- **Compare and contrast** the crystals that form from salt and sugar solutions.
- **Observe** crystals and infer how they formed.

Materials
250-mL beakers (2)   cotton string
cardboard   hot plate
large paper clip   magnifying lens
table salt   thermal mitt
flat wooden stick   shallow pan
granulated sugar   spoon

Safety Precautions

WARNING: *Never taste or eat any lab materials.*

● Procedure

1. Gently mix separate solutions of salt in water and sugar in water in the two beakers. Keep stirring the solutions as you add salt or sugar to the water. Stop mixing when no more salt or sugar will dissolve in the solutions. Label each beaker.

2. Place the sugar solution beaker on a hot plate. Use the hot plate to heat the sugar solution gently. **WARNING:** *Do not touch the hot beaker without protecting your hands.*

3. Tie one end of the thread to the middle of the wooden stick. Tie a large paper clip to the free end of the string for weight. Place the stick across the opening of the sugar beaker so the thread dangles in the sugar solution.

4. Remove the beaker from the hot plate and cover it with cardboard. Place it in a location where it won't be disturbed.

5. Pour a thin layer of the salt solution into the shallow pan.

6. Leave the beaker and the shallow pan undisturbed for at least one week.

7. After one week, examine each solution with a magnifying lens to see whether crystals have formed.

● Conclude and Apply

1. **Compare and contrast** the crystals that formed from the salt and the sugar solutions. How do they compare with samples of table salt and sugar?
2. **Describe** what happened to the saltwater solution in the shallow pan.
3. Did this same process occur in the sugar solution? Explain.

*C*ommunicating Your Data

Make a poster that describes your methods of growing salt and sugar crystals. Present your results to your class.

LAB F ◆ 13

section 2
Mineral Identification

as you read

What You'll Learn
- **Describe** physical properties used to identify minerals.
- **Identify** minerals using physical properties such as hardness and streak.

Why It's Important
Identifying minerals helps you recognize valuable mineral resources.

Review Vocabulary
physical property: any characteristic of a material that you can observe without changing the identity of the material

New Vocabulary
- hardness
- luster
- specific gravity
- streak
- cleavage
- fracture

Physical Properties

Why can you recognize a classmate when you see him or her in a crowd away from school? A person's height or the shape of his or her face helps you tell that person from the rest of your class. Height and facial shape are two properties unique to individuals. Individual minerals also have unique properties that distinguish them.

Mineral Appearance Just like height and facial characteristics help you recognize someone, mineral properties can help you recognize and distinguish minerals. Color and appearance are two obvious clues that can be used to identify minerals.

However, these clues alone aren't enough to recognize most minerals. The minerals pyrite and gold are gold in color and can appear similar, as shown in **Figure 6.** As a matter of fact, pyrite often is called fool's gold. Gold is worth a lot of money, whereas pyrite has little value. You need to look at other properties of minerals to tell them apart. Some other properties to study include how hard a mineral is, how it breaks, and its color when crushed into a powder. Every property you observe in a mineral is a clue to its identity.

Figure 6 The general appearance of a mineral often is not enough to identify it.

Pyrite

Gold

Using only color, observers can be fooled when trying to distinguish between pyrite and gold.

Azurite

The mineral azurite is identified readily by its striking blue color.

Hardness A measure of how easily a mineral can be scratched is its **hardness.** The mineral talc is so soft you can scratch it loose with your fingernail. Talcum powder is made from this soft mineral. Diamonds, on the other hand, are the hardest mineral. Some diamonds are used as cutting tools, as shown in **Figure 7.** A diamond can be scratched only by another diamond. Diamonds can be broken, however.

 Why is hardness sometimes referred to as scratchability?

Sometimes the concept of hardness is confused with whether or not a mineral will break. It is important to understand that even though a diamond is extremely hard, it can shatter if given a hard enough blow in the right direction along the crystal.

Mohs Scale In 1824, the Austrian scientist Friedrich Mohs developed a list of common minerals to compare their hardnesses. This list is called Mohs scale of hardness, as seen in **Table 1.** The scale lists the hardness of ten minerals. Talc, the softest mineral, has a hardness value of one, and diamond, the hardest mineral, has a value of ten.

Here's how the scale works. Imagine that you have a clear or whitish-colored mineral that you know is either fluorite or quartz. You try to scratch it with your fingernail and then with an iron nail. You can't scratch it with your fingernail but you can scratch it with the iron nail. Because the hardness of your fingernail is 2.5 and that of the iron nail is 4.5, you can determine the unknown mineral's hardness to be somewhere around 3 or 4. Because it is known that quartz has a hardness of 7 and fluorite has a hardness of 4, the mystery mineral must be fluorite.

Some minerals have a hardness range rather than a single hardness value. This is because atoms are arranged differently in different directions in their crystal structures.

Figure 7 Some saw blades have diamonds embedded in them to help slice through materials, such as this limestone. Blades are kept cool by running water over them.

Table 1 Mineral Hardness		
Mohs Scale	Hardness	Hardness of Common Objects
Talc (softest)	1	
Gypsum	2	fingernail (2.5)
Calcite	3	piece of copper (2.5 to 3.0)
Fluorite	4	iron nail (4.5)
Apatite	5	glass (5.5)
Feldspar	6	steel file (6.5)
Quartz	7	streak plate (7.0)
Topaz	8	
Corundum	9	
Diamond (hardest)	10	

Graphite

Fluorite

Figure 8 Luster is an important physical property that is used to distinguish minerals. Graphite has a metallic luster. Fluorite has a nonmetallic, glassy luster.

Luster The way a mineral reflects light is known as **luster.** Luster can be metallic or nonmetallic. Minerals with a metallic luster, like the graphite shown in **Figure 8,** shine like metal. Metallic luster can be compared to the shine of a metal belt buckle, the shiny chrome trim on some cars, or the shine of metallic cooking utensils. When a mineral does not shine like metal, its luster is nonmetallic. Examples of terms for nonmetallic luster include dull, pearly, silky, and glassy. Common examples of minerals with glassy luster are quartz, calcite, halite, and fluorite.

Specific Gravity Minerals also can be distinguished by comparing the weights of equal-sized samples. The **specific gravity** of a mineral is the ratio of its weight compared with the weight of an equal volume of water. Like hardness, specific gravity is expressed as a number. If you were to research the specific gravities of gold and pyrite, you'd find that gold's specific gravity is about 19, and pyrite's is 5. This means that gold is about 19 times heavier than water and pyrite is 5 times heavier than water. You could experience this by comparing equal-sized samples of gold and pyrite in your hands—the pyrite would feel much lighter. The term *heft* is sometimes used to describe how heavy a mineral sample feels.

Applying Science

How can you identify minerals?

Properties of Minerals

Mineral	Hardness	Streak
Copper	2.5–3	copper-red
Galena	2.5	dark gray
Gold	2.5–3	yellow
Hematite	5.5–6.5	red to brown
Magnetite	6–6.5	black
Silver	2.5–3	silver-white

You have learned that minerals are identified by their physical properties, such as streak, hardness, cleavage, and color. Use your knowledge of mineral properties and your ability to read a table to solve the following problems.

Identifying the Problem
The table includes hardnesses and streak colors for several minerals. How can you use these data to distinguish minerals?

Solving the Problem
1. What test would you perform to distinguish hematite from copper? How would you carry out this test?
2. How could you distinguish copper from galena? What tool would you use?
3. What would you do if two minerals had the same hardness and the same streak color?

Streak When a mineral is rubbed across a piece of unglazed porcelain tile, as in **Figure 9,** a streak of powdered mineral is left behind. **Streak** is the color of a mineral when it is in a powdered form. The streak test works only for minerals that are softer than the streak plate. Gold and pyrite can be distinguished by a streak test. Gold has a yellow streak and pyrite has a greenish-black or brownish-black streak.

Some soft minerals will leave a streak even on paper. The last time you used a pencil to write on paper, you left a streak of the mineral graphite. One reason that graphite is used in pencil lead is because it is soft enough to leave a streak on paper.

 Why do gold and pyrite leave a streak, but quartz does not?

Figure 9 Streak is more useful for mineral identification than is mineral color. Hematite, for example, can be dark red, gray, or silver in color. However, its streak is always dark reddish-brown.

Cleavage and Fracture The way a mineral breaks is another clue to its identity. Minerals that break along smooth, flat surfaces have **cleavage** (KLEE vihj). Cleavage, like hardness, is determined partly by the arrangement of the mineral's atoms. Mica is a mineral that has one perfect cleavage. **Figure 10** shows how mica breaks along smooth, flat planes. If you were to take a layer cake and separate its layers, you would show that the cake has cleavage. Not all minerals have cleavage. Minerals that break with uneven, rough, or jagged surfaces have **fracture.** Quartz is a mineral with fracture. If you were to grab a chunk out of the side of that cake, it would be like breaking a mineral that has fracture.

Mica

Halite

Figure 10 Weak or fewer bonds within the structures of mica and halite allow them to be broken along smooth, flat cleavage planes. **Infer** *If you broke quartz, would it look the same?*

Mini LAB

Observing Mineral Properties

Procedure
1. Obtain samples of some of the following clear minerals: **gypsum, muscovite mica, halite,** and **calcite.**
2. Place each sample over the print on this page and observe the letters.

Analysis
1. Which mineral can be identified by observing the print's double image?
2. What other special property is used to identify this mineral?

Figure 11 Some minerals are natural magnets, such as this lodestone, which is a variety of magnetite.

Other Properties Some minerals have unique properties. Magnetite, as you can guess by its name, is attracted to magnets. Lodestone, a form of magnetite, will pick up iron filings like a magnet, as shown in **Figure 11.** Light forms two separate rays when it passes through calcite, causing you to see a double image when viewed through transparent specimens. Calcite also can be identified because it fizzes when hydrochloric acid is put on it.

Now you know that you sometimes need more information than color and appearance to identify a mineral. You also might need to test its streak, hardness, luster, and cleavage or fracture. Although the overall appearance of a mineral can be different from sample to sample, its physical properties remain the same.

section 2 review

Summary

Physical Properties
- Minerals are identified by observing their physical properties.
- Hardness is a measure of how easily a mineral can be scratched.
- Luster describes how a mineral reflects light.
- Specific gravity is the ratio of the weight of a mineral sample compared to the weight of an equal volume of water.
- Streak is the color of a powdered mineral.
- Minerals with cleavage break along smooth, flat surfaces in one or more directions.
- Fracture describes any uneven manner in which a mineral breaks.
- Some minerals react readily with acid, form a double image, or are magnetic.

Self Check

1. **Compare and contrast** a mineral fragment that has one cleavage direction with one that has only fracture.
2. **Explain** how an unglazed porcelain tile can be used to identify a mineral.
3. **Explain** why streak often is more useful for mineral identification than color.
4. **Determine** What hardness does a mineral have if it does not scratch glass but it scratches an iron nail?
5. **Think Critically** What does the presence of cleavage planes within a mineral tell you about the chemical bonds that hold the mineral together?

Applying Skills

6. **Draw Conclusions** A large piece of the mineral halite is broken repeatedly into several perfect cubes. How can this be explained?

bookf.msscience.com/self_check_quiz

section 3
Uses of Minerals

Gems

Walking past the window of a jewelry store, you notice a large selection of beautiful jewelry—a watch sparkling with diamonds, a necklace holding a brilliant red ruby, and a gold ring. For thousands of years, people have worn and prized minerals in their jewelry. What makes some minerals special? What unusual properties do they have that make them so valuable?

Properties of Gems As you can see in **Figure 12, gems** or gemstones are highly prized minerals because they are rare and beautiful. Most gems are special varieties of a particular mineral. They are clearer, brighter, or more colorful than common samples of that mineral. The difference between a gem and the common form of the same mineral can be slight. Amethyst is a gem form of quartz that contains just traces of iron in its structure. This small amount of iron gives amethyst a desirable purple color. Sometimes a gem has a crystal structure that allows it to be cut and polished to a higher quality than that of a non-gem mineral. **Table 2** lists popular gems and some locations where they have been collected.

as you read

What You'll Learn
- **Describe** characteristics of gems that make them more valuable than other minerals.
- **Identify** useful elements that are contained in minerals.

Why It's Important

Minerals are necessary materials for decorative items and many manufactured products.

Review Vocabulary
metal: element that typically is a shiny, malleable solid that conducts heat and electricity well

New Vocabulary
- gem
- ore

Figure 12 It is easy to see why gems are prized for their beauty and rarity. Shown here is The Imperial State Crown, made for Queen Victoria of England in 1838. It contains thousands of jewels, including diamonds, rubies, sapphires, and emeralds.

SECTION 3 Uses of Minerals **F** ◆ **19**

Table 2 Minerals and Their Gems

Fun Facts	Mineral	Gem Example	Some Important Locations
Beryl is named for the element beryllium, which it contains. Some crystals reach several meters in length.	Beryl	Emerald	Colombia, Brazil, South Africa, North Carolina
A red spinel in the British crown jewels has a mass of 352 carats. A carat is 0.2 g.	Spinel	Ruby spinel	Sri Lanka, Thailand, Myanmar (Burma)
Purplish-blue examples of zoisite were discovered in 1967 near Arusha, Tanzania.	Zoisite	Tanzanite	Tanzania
The most valuable examples are yellow, pink, and blue varieties.	Topaz (uncut)	Topaz (gem)	Siberia, Germany, Japan, Mexico, Brazil, Colorado, Utah, Texas, California, Maine, Virginia, South Carolina

Fun Facts	Mineral	Gem Example	Some Important Locations
Olivine composes a large part of Earth's upper mantle. It is also present in moon rocks.	Olivine	Peridot	Myanmar (Burma), Zebirget (Saint John's Island, located in the Red Sea), Arizona, New Mexico
Garnet is a common mineral found in a wide variety of rock types. The red color of the variety almandine is caused by iron in its crystal structure.	Garnet	Almandine	Ural Mountains, Italy, Madagascar, Czech Republic, India, Sri Lanka, Brazil, North Carolina, Arizona, New Mexico
Quartz makes up about 30 percent of Earth's continental crust.	Quartz	Amethyst	Colorless varieties in Hot Springs, Arkansas; Amethyst in Brazil, Uruguay, Madagascar, Montana, North Carolina, California, Maine
The blue color of sapphire is caused by iron or titanium in corundum. Chromium in corundum produces the red color of ruby.	Corundum	Blue sapphire	Thailand, Cambodia, Sri Lanka, Kashmir

Important Gems All gems are prized, but some are truly spectacular and have played an important role in history. For example, the Cullinan diamond, found in South Africa in 1905, was the largest uncut diamond ever discovered. Its mass was 3,106.75 carats (about 621 g). The Cullinan diamond was cut into 9 main stones and 96 smaller ones. The largest of these is called the Cullinan 1 or Great Star of Africa. Its mass is 530.20 carats (about 106 g), and it is now part of the British monarchy's crown jewels, shown in **Figure 13A.**

Another well-known diamond is the blue Hope diamond, shown in **Figure 13B.** This is perhaps the most notorious of all diamonds. It was purchased by Henry Philip Hope around 1830, after whom it is named. Because his entire family as well as a later owner suffered misfortune, the Hope diamond has gained a reputation for bringing its owner bad luck. The Hope diamond's mass is 45.52 carats (about 9 g). Currently it is displayed in the Smithsonian Institution in Washington, D.C.

Useful Gems In addition to their beauty, some gems serve useful purposes. You learned earlier that diamonds have a hardness of 10 on Mohs scale. They can scratch almost any material—a property that makes them useful as industrial abrasives and cutting tools. Other useful gems include rubies, which are used to produce specific types of laser light. Quartz crystals are used in electronics and as timepieces. When subjected to an electric field, quartz vibrates steadily, which helps control frequencies in electronic devices and allows for accurate timekeeping.

Most industrial diamonds and other gems are synthetic, which means that humans make them. However, the study of natural gems led to their synthesis, allowing the synthetic varieties to be used by humans readily.

Topic: Gemstone Data
Visit bookf.msscience.com for Web links to information about gems at the Smithsonian Museum of Natural History.

Activity List three important examples of gems other than those described on this page. Prepare a data table with the heads *Gem Name/Type, Weight (carats/grams), Mineral,* and *Location*. Fill in the table entries for the gemstones you selected.

Figure 13 These gems are among the most famous examples of precious stones.

A The Great Star of Africa is part of a sceptre in the collection of British crown jewels.

B Beginning in 1668, the Hope diamond was part of the French crown jewels. Then known as the French Blue, it was stolen in 1792 and later surfaced in London, England in 1812.

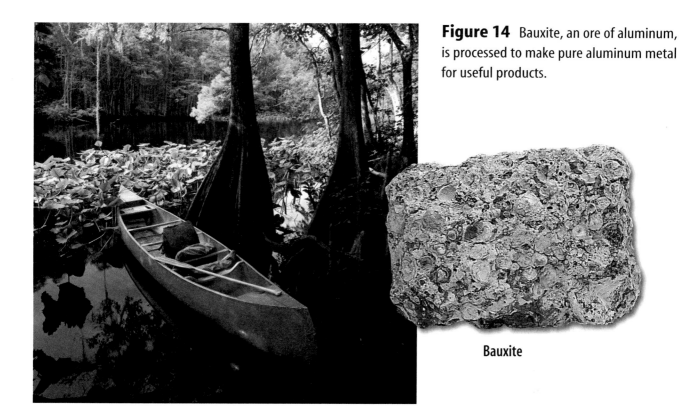

Figure 14 Bauxite, an ore of aluminum, is processed to make pure aluminum metal for useful products.

Bauxite

Useful Elements in Minerals

Gemstones are perhaps the best-known use of minerals, but they are not the most important. Look around your home. How many things made from minerals can you name? Can you find anything made from iron?

Ores Iron, used in everything from frying pans to ships, is obtained from its ore, hematite. A mineral or rock is an **ore** if it contains a useful substance that can be mined at a profit. Magnetite is another mineral that contains iron.

Reading Check *When is a mineral also an ore?*

 Aluminum sometimes is refined, or purified, from the ore bauxite, shown in **Figure 14.** In the process of refining aluminum, aluminum oxide powder is separated from unwanted materials that are present in the original bauxite. After this, the aluminum oxide powder is converted to molten aluminum by a process called smelting.

During smelting, a substance is melted to separate it from any unwanted materials that may remain. Aluminum can be made into useful products like bicycles, soft-drink cans, foil, and lightweight parts for airplanes and cars. The plane flown by the Wright brothers during the first flight at Kitty Hawk had an engine made partly of aluminum.

Historical Mineralogy An early scientific description of minerals was published by Georgius Agricola in 1556. Use print and online resources to research the mining techniques discussed by Agricola in his work *De Re Metallica*.

Figure 15 The mineral sphalerite (greenish when nearly pure) is an important source of zinc. Iron often is coated with zinc to prevent rust in a process called galvanization.

Vein Minerals Under certain conditions, metallic elements can dissolve in fluids. These fluids then travel through weaknesses in rocks and form mineral deposits. Weaknesses in rocks include natural fractures or cracks, faults, and surfaces between layered rock formations. Mineral deposits left behind that fill in the open spaces created by the weaknesses are called vein mineral deposits.

Reading Check *How do fluids move through rocks?*

Sometimes vein mineral deposits fill in the empty spaces after rocks collapse. An example of a mineral that can form in this way is shown in **Figure 15.** This is the shiny mineral sphalerite, a source of the element zinc, which is used in batteries. Sphalerite sometimes fills spaces in collapsed limestone.

Minerals Containing Titanium You might own golf clubs with titanium shafts or a racing bicycle containing titanium. Perhaps you know someone who has a titanium hip or knee replacement. Titanium is a durable, lightweight, metallic element derived from minerals that contain this metal in their crystal structures. Two minerals that are sources of the element titanium are ilmenite (IHL muh nite) and rutile (rew TEEL), shown in **Figure 16.** Ilmenite and rutile are common in rocks that form when magma cools and solidifies. They also occur as vein mineral deposits and in beach sands.

Figure 16 Rutile and ilmenite are common ore minerals of the element titanium.

Rutile

Ilmenite

Figure 17 Wheelchairs used for racing and playing basketball often have parts made from titanium.

Uses for Titanium Titanium is used in automobile body parts, such as connecting rods, valves, and suspension springs. Low density and durability make it useful in the manufacture of aircraft, eyeglass frames, and sports equipment such as tennis rackets and bicycles. Wheelchairs used by people who want to race or play basketball often are made from titanium, as shown in **Figure 17.** Titanium is one of many examples of useful materials that come from minerals and that enrich humans' lives.

section 3 review

Summary

Gems
- Gems are highly prized mineral specimens often used as decorative pieces in jewelry or other items.
- Some gems, especially synthetic ones, have industrial uses.

Useful Elements in Minerals
- Economically important quantities of useful elements or compounds are present in ores.
- Ores generally must be processed to extract the desired material.
- Iron, aluminum, zinc, and titanium are common metals that are extracted from minerals.

Self Check

1. **Explain** why the Cullinan diamond is an important gem.
2. **Identify** Examine **Table 2.** What do rubies and sapphires have in common?
3. **Describe** how vein minerals form.
4. **Explain** why bauxite is considered to be a useful rock.
5. **Think Critically** Titanium is nontoxic. Why is this important in the manufacture of artificial body parts?

Applying Skills

6. **Use Percentages** Earth's average continental crust contains 5 percent iron and 0.007 percent zinc. How many times more iron than zinc is present in average continental crust?

Mineral Identification

Real-World Question
Although certain minerals can be identified by observing only one property, others require testing several properties to identify them. How can you identify unknown minerals?

Goals
- **Hypothesize** which properties of each mineral are most useful for identification purposes.
- **Test** your hypothesis as you attempt to identify unknown mineral samples.

Materials
mineral samples
magnifying lens
pan balance
graduated cylinder
water
piece of copper
copper penny
glass plate
small iron nail
steel file
streak plate
5% HCl with dropper
Mohs scale of hardness
Minerals Appendix
minerals field guide
safety goggles
*Alternate materials

Safety Precautions

WARNING: *If an HCl spill occurs, notify your teacher and rinse with cool water until you are told to stop. Do not taste, eat, or drink any lab materials.*

Procedure
1. Copy the data table into your Science Journal. Obtain a set of unknown minerals.
2. Observe a numbered mineral specimen carefully. Write a star in the table entry that represents what you hypothesize is an important physical property. Choose one or two properties that you think will help most in identifying the sample.
3. Perform tests to observe your chosen properties first.
   a. To estimate hardness:
      - Rub the sample firmly against objects of known hardness and observe whether it leaves a scratch on the objects.
      - Estimate a hardness range based on which items the mineral scratches.
   b. To estimate specific gravity: Perform a density measurement.
      - Use the pan balance to determine the sample's mass, in grams.

Using Scientific Methods

- Measure its volume using a graduated cylinder partially filled with water. The amount of water displaced by the immersed sample, in mL, is an estimate of its volume in cm³.
- Divide mass by volume to determine density. This number, without units, is comparable to specific gravity.

4. With the help of the Mineral Appendix or a field guide, attempt to identify the sample using the properties from step 2. Perform more physical property observations until you can identify the sample. Repeat steps 2 through 4 for each unknown.

Physical Properties of Minerals

Sample Number	Hardness	Cleavage or Fracture	Color	Specific Gravity	Luster and Streak	Crystal Shape	Other Properties	Mineral Name
1								
2				Do not write in this book.				
etc.								

Analyze Your Data

1. Which properties were most useful in identifying your samples? Which properties were least useful?
2. **Compare** the properties that worked best for you with those that worked best for other students.

Conclude and Apply

1. **Determine** two properties that distinguish clear, transparent quartz from clear, transparent calcite. Explain your choice of properties.
2. Which physical properties would be easiest to determine if you found a mineral specimen in the field?

Communicating Your Data

For three minerals, list physical properties that were important for their identification. **For more help, refer to the** Science Skill Handbook.

TIME SCIENCE AND HISTORY

SCIENCE CAN CHANGE THE COURSE OF HISTORY!

Dr. Dorothy Crowfoot Hodgkin

Like X rays, electrons are diffracted by crystalline substances, revealing information about their internal structures and symmetry. This electron diffraction pattern of titanium was obtained with an electron beam focused along a specific direction in the crystal.

Trailblazing scientist and humanitarian

What contributions did Dorothy Crowfoot Hodgkin make to science?

Dr. Hodgkin used a method called X-ray crystallography (kris tuh LAH gruh fee) to figure out the structures of crystalline substances, including vitamin B^{12}, vitamin D, penicillin, and insulin.

What's X-ray crystallography?

Scientists expose a crystalline sample to X rays. As X rays travel through a crystal, the crystal diffracts, or scatters, the X rays into a regular pattern. Like an individual's fingerprints, each crystalline substance has a unique diffraction pattern. Crystallography has applications in the life, Earth, and physical sciences. For example, geologists use X-ray crystallography to identify and study minerals found in rocks.

What were some obstacles Hodgkin overcame?

During the 1930s, there were few women scientists. Hodgkin was not even allowed to attend meetings of the chemistry faculty where she taught because she was a woman. Eventually, she won over her colleagues with her intelligence and tenacity.

How does Hodgkin's research help people today?

Dr. Hodgkin's discovery of the structure of insulin helped scientists learn how to control diabetes, a disease that affects more than 15 million Americans. Diabetics' bodies are unable to process sugar efficiently. Diabetes can be fatal. Fortunately, Dr. Hodgkin's research with insulin has saved many lives.

1910–1994

Research Look in reference books or go to the Glencoe Science Web site for information on how X-ray crystallography is used to study minerals. Write your findings and share them with your class.

For more information, visit bookf.msscience.com/time

Chapter 1 Study Guide

Reviewing Main Ideas

Section 1 Minerals

1. Much of what you use each day is made at least in some part from minerals.
2. All minerals are formed by natural processes and are inorganic solids with definite chemical compositions and orderly arrangements of atoms.
3. Minerals have crystal structures in one of six major crystal systems.

Section 2 Mineral Identification

1. Hardness is a measure of how easily a mineral can be scratched.
2. Luster describes how light reflects from a mineral's surface.
3. Streak is the color of the powder left by a mineral on an unglazed porcelain tile.
4. Minerals that break along smooth, flat surfaces have cleavage. When minerals break with rough or jagged surfaces, they are displaying fracture.
5. Some minerals have special properties that aid in identifying them. For example, magnetite is identified by its attraction to a magnet.

Section 3 Uses of Minerals

1. Gems are minerals that are more rare and beautiful than common minerals.
2. Minerals are useful for their physical properties and for the elements they contain.

Visualizing Main Ideas

Copy and complete the following concept map about minerals. Use the following words and phrases: the way a mineral breaks, the way a mineral reflects light, ore, a rare and beautiful mineral, how easily a mineral is scratched, streak, *and* a useful substance mined for profit.

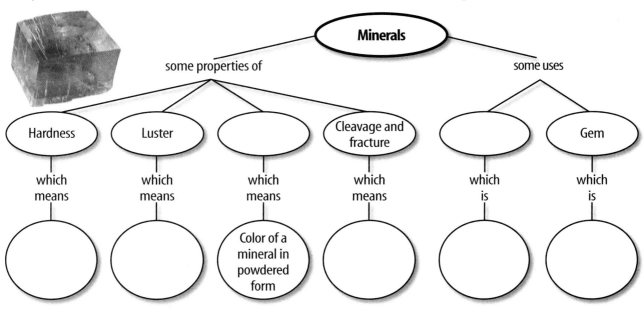

chapter 1 Review

Using Vocabulary

cleavage p. 17
crystal p. 9
fracture p. 17
gem p. 19
hardness p. 15
luster p. 16
magma p. 11
mineral p. 8
ore p. 23
silicate p. 12
specific gravity p. 16
streak p. 17

Explain the difference between the vocabulary words in each of the following sets.

1. cleavage—fracture
2. crystal—mineral
3. luster—streak
4. magma—crystal
5. hardness—specific gravity
6. ore—mineral
7. crystal—luster
8. mineral—silicate
9. gem—crystal
10. streak—specific gravity

Checking Concepts

Choose the word or phrase that best answers the question.

11. Which is a characteristic of a mineral?
    A) It can be a liquid.
    B) It is organic.
    C) It has no crystal structure.
    D) It is inorganic.

12. What must all silicates contain?
    A) magnesium
    B) silicon and oxygen
    C) silicon and aluminum
    D) oxygen and carbon

13. What is the measure of how easily a mineral can be scratched?
    A) luster
    B) hardness
    C) cleavage
    D) fracture

Use the photo below to answer question 14.

14. Examine the photo of quartz above. In what way does quartz break?
    A) cleavage
    B) fracture
    C) luster
    D) flat planes

15. Which of the following must crystalline solids have?
    A) carbonates
    B) cubic structures
    C) orderly arrangement of atoms
    D) cleavage

16. What is the color of a powdered mineral formed when rubbing it against an unglazed porcelain tile?
    A) luster
    B) density
    C) hardness
    D) streak

17. Which is hardest on Mohs scale?
    A) talc
    B) quartz
    C) diamond
    D) feldspar

30 ◆ F CHAPTER REVIEW   bookf.msscience.com/vocabulary_puzzlemaker

chapter 1 Review

Thinking Critically

18. **Classify** Water is an inorganic substance that is formed by natural processes on Earth. It has a unique composition. Sometimes water is a mineral and other times it is not. Explain.

19. **Determine** how many sides a perfect salt crystal has.

20. **Apply** Suppose you let a sugar solution evaporate, leaving sugar crystals behind. Are these crystals minerals? Explain.

21. **Predict** Will a diamond leave a streak on a streak plate? Explain.

22. **Collect Data** Make an outline of how at least seven physical properties can be used to identify unknown minerals.

23. **Explain** how you would use **Table 1** to determine the hardness of any mineral.

24. **Concept Map** Copy and complete the concept map below, which includes two crystal systems and two examples from each system. Use the following words and phrases: *hexagonal, corundum, halite, fluorite,* and *quartz*.

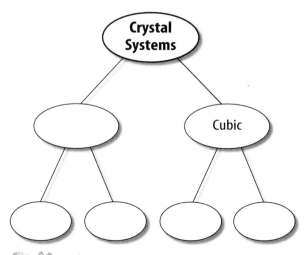

Performance Activities

25. **Display** Make a display that shows the six crystal systems of minerals. Research the crystal systems of minerals and give three examples for each crystal system. Indicate whether any of the minerals are found in your state. Describe any important uses of these minerals. Present your display to the class.

Applying Math

26. **Mineral Volume** Recall that 1 mL = 1 cm³. Suppose that the volume of water in a graduated cylinder is 107.5 mL. A specimen of quartz, tied to a piece of string, is immersed in the water. The new water level reads 186 mL. What is the volume, in cm³, of the piece of quartz?

Use the graph below to answer questions 27 and 28.

27. **Zinc Use** According to the graph above, what was the main use of zinc consumed in the United States between 1978 and 1998?

28. **Metal Products** According to the graph, approximately how many thousand metric tons of zinc were used to make brass and bronze products in 1998?

Chapter 1 Standardized Test Practice

Part 1 Multiple Choice

Record your answers on the answer sheet provided by your teacher or on a sheet of paper.

Use the photo below to answer question 1.

1. To which crystal system does the crystal shown above belong?
   A. hexagonal C. triclinic
   B. cubic   D. monoclinic

2. Which of the following is a common rock-forming mineral?
   A. azurite   C. quartz
   B. gold   D. diamond

3. Which term refers to the resistance of a mineral to scratching?
   A. hardness   C. luster
   B. specific gravity D. fracture

4. Which is a special property of the mineral magnetite?
   A. attracted by a magnet
   B. fizzes with dilute hydrochloric acid
   C. forms a double image
   D. has a salty taste

5. Which causes some minerals to break along smooth, flat surfaces?
   A. streak   C. luster
   B. cleavage D. fracture

Test-Taking Tip

If you are taking a timed test, keep track of time during the test. If you find that you're spending too much time on a multiple-choice question, mark your best guess and move on.

6. Which of these forms in cracks or along faults?
   A. bauxite
   B. silicates
   C. vein minerals
   D. rock-forming minerals

7. Which is the most abundant element in Earth's crust?
   A. silicon   C. iron
   B. manganese   D. oxygen

Use the table below to answer questions 8–10.

Mineral	Hardness
Talc	1
Gypsum	2
Calcite	3
Fluorite	4
Apatite	5
Feldspar	6
Quartz	7
Topaz	8
Corundum	9
Diamond	10

8. Which mineral in the table is softest?
   A. diamond   C. talc
   B. feldspar D. gypsum

9. Which mineral will scratch feldspar but not topaz?
   A. quartz   C. apatite
   B. calcite   D. diamond

10. After whom is the scale shown above named?
    A. Neil Armstrong
    B. Friedrich Mohs
    C. Alfred Wegener
    D. Isaac Newton

Standardized Test Practice

Part 2 | Short Response/Grid In

Record your answers on the answer sheet provided by your teacher or on a sheet of paper.

11. What is the definition of a mineral?
12. Why are gems valuable?
13. Explain the difference between fracture and cleavage.
14. Why is mineral color sometimes not helpful for identifying minerals?

Use the conversion factor and table below to answer questions 15–17.

1.0 carat = 0.2 grams

Diamond	Carats	Grams
Uncle Sam: largest diamond found in United States	40.4	?
Punch Jones: second largest U.S. diamond; named after boy who discovered it	?	6.89
Theresa: discovered in Wisconsin in 1888	21.5	4.3
2001 diamond production from western Australia	21,679,930	?

15. How many grams is the *Uncle Sam* diamond?
16. How many carats is the *Punch Jones* diamond?
17. How many grams of diamond were produced in western Australia in 2001?
18. What is the source of most of the diamonds that are used for industrial purposes?
19. Explain how minerals are useful to society. Describe some of their uses.

Part 3 | Open Ended

Record your answers on a sheet of paper.

Use the photo below to answer question 20.

20. The mineral crystals in the rock above formed when magma cooled and are visible with the unaided eye. Hypothesize about how fast the magma cooled.
21. What is a crystal system? Why is it useful to classify mineral crystals this way?
22. How can a mineral be identified using its physical properties?
23. What is a crystal? Do all crystals have smooth crystal faces? Explain.
24. Are gases that are given off by volcanoes minerals? Why or why not?
25. What is the most abundant mineral group in Earth's crust? What elements always are found in the minerals included in this group?
26. Several layers are peeled from a piece of muscovite mica? What property of minerals does this illustrate? Describe this property in mica.

chapter 2

Rocks

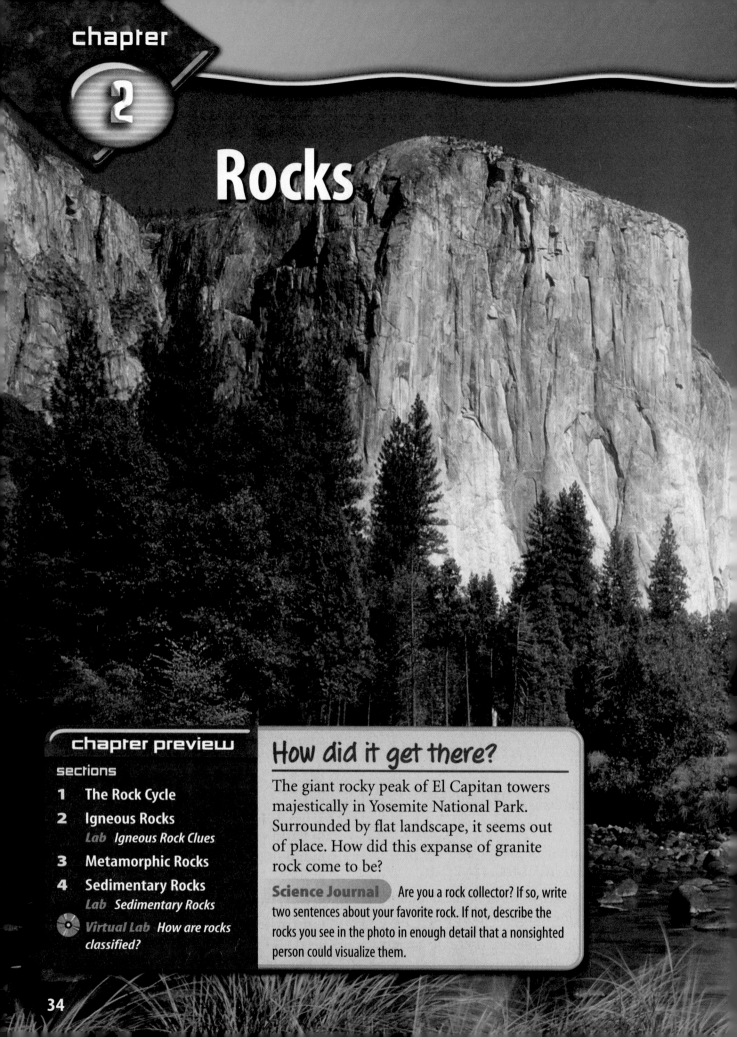

chapter preview

sections

1 **The Rock Cycle**
2 **Igneous Rocks**
   Lab Igneous Rock Clues
3 **Metamorphic Rocks**
4 **Sedimentary Rocks**
   Lab Sedimentary Rocks
   Virtual Lab How are rocks classified?

How did it get there?

The giant rocky peak of El Capitan towers majestically in Yosemite National Park. Surrounded by flat landscape, it seems out of place. How did this expanse of granite rock come to be?

Science Journal Are you a rock collector? If so, write two sentences about your favorite rock. If not, describe the rocks you see in the photo in enough detail that a nonsighted person could visualize them.

Start-Up Activities

Observe and Describe Rocks

Some rocks are made of small mineral grains that lock together, like pieces of a puzzle. Others are grains of sand tightly held together or solidified lava that once flowed from a volcano. If you examine rocks closely, you sometimes can tell what they are made of.

1. Collect three different rock samples near your home or school.
2. Draw a picture of the details you see in each rock.
3. Use a magnifying lens to look for different types of materials within the same rock.
4. Describe the characteristics of each rock. Compare your drawings and descriptions with photos, drawings, and descriptions in a rocks and minerals field guide.
5. Use the field guide to try to identify each rock.
6. **Think Critically** Decide whether you think your rocks are mixtures. If so, infer or suggest what these mixtures might contain. Write your explanations in your Science Journal.

Major Rock Types Make the following Foldable to help you organize facts about types of rocks.

STEP 1 Fold a sheet of paper in half lengthwise. Make the back edge about 5 cm longer than the front edge.

STEP 2 Turn the paper so the fold is on the bottom. Then fold it into thirds.

STEP 3 Unfold and cut only the top layer along both folds to make three tabs.

STEP 4 Label the Foldable as shown.

Make an Organizational Study Fold As you read the chapter, write and illustrate what you learn about the three main types of rocks in your study fold.

Preview this chapter's content and activities at bookf.msscience.com

section 1
The Rock Cycle

as you read

What You'll Learn
- **Distinguish** between a rock and a mineral.
- **Describe** the rock cycle and some changes that a rock could undergo.

Why It's Important
Rocks exist everywhere, from under deep oceans and in high mountain ranges, to the landscape beneath your feet.

Review Vocabulary
mineral: a naturally occurring, inorganic solid with a definite chemical composition and an orderly arrangement of atoms

New Vocabulary
- rock
- rock cycle

What is a rock?

Imagine you and some friends are exploring a creek. Your eye catches a glint from a piece of rock at the edge of the water. As you wander over to pick up the rock, you notice that it is made of different-colored materials. Some of the colors reflect light, while others are dull. You put the rock in your pocket for closer inspection in science lab.

Common Rocks The next time you walk past a large building or monument, stop and take a close look at it. Chances are that it is made out of common rock. In fact, most rock used for building stone contains one or more common minerals, called rock-forming minerals, such as quartz, feldspar, mica, or calcite. When you look closely, the sparkles you see are individual crystals of minerals. A **rock** is a mixture of such minerals, rock fragments, volcanic glass, organic matter, or other natural materials. **Figure 1** shows minerals mixed together to form the rock granite. You might even find granite near your home.

Figure 1 Mount Rushmore, in South Dakota, is made of granite. Granite is a mixture of feldspar, quartz, mica, hornblende, and other minerals.

36 ◆ F CHAPTER 2 Rocks

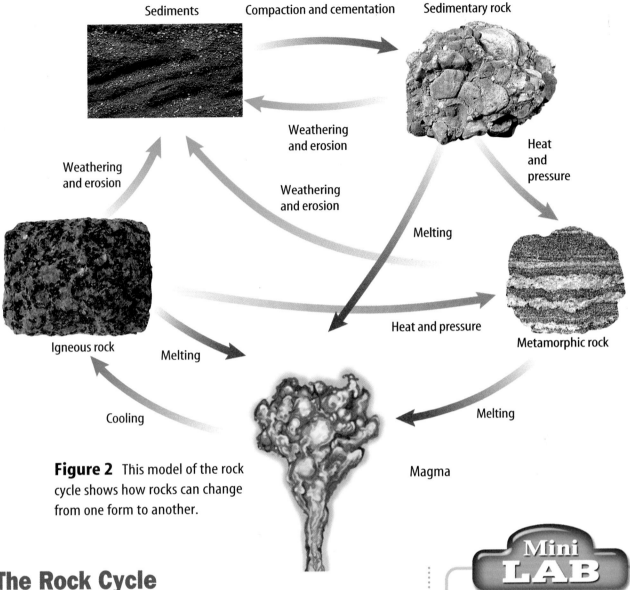

Figure 2 This model of the rock cycle shows how rocks can change from one form to another.

The Rock Cycle

To show how rocks slowly change through time, scientists have created a model called the **rock cycle,** shown in **Figure 2.** It illustrates the processes that create and change rocks. The rock cycle shows the three types of rock—igneous, metamorphic, and sedimentary—and the processes that form them.

Look at the rock cycle and notice that rocks change by many processes. For example, a sedimentary rock can change by heat and pressure to form a metamorphic rock. The metamorphic rock then can melt and later cool to form an igneous rock. The igneous rock then could be broken into fragments by weathering and erode away. The fragments might later compact and cement together to form another sedimentary rock. Any given rock can change into any of the three major rock types. A rock even can transform into another rock of the same type.

Reading Check *What is illustrated by the rock cycle?*

Mini LAB

Modeling Rock

Procedure
1. Mix about 10 mL of **white glue** with about 7 g of **dirt** or **sand** in a **small paper cup.**
2. Stir the mixture and then allow it to harden overnight.
3. Tear away the paper cup carefully from your mixture.

Analysis
1. Which rock type is similar to your hardened mixture?
2. Which part of the rock cycle did you model?

Try at Home

NATIONAL GEOGRAPHIC VISUALIZING THE ROCK CYCLE

Figure 3

Rocks continuously form and transform in a process that geologists call the rock cycle. For example, molten rock—from volcanoes such as Washington's Mount Rainier, background—cools and solidifies to form igneous rock. It slowly breaks down when exposed to air and water to form sediments. These sediments are compacted or cemented into sedimentary rock. Heat and pressure might transform sedimentary rock into metamorphic rock. When metamorphic rock melts and hardens, igneous rock forms again. There is no distinct beginning, nor is there an end, to the rock cycle.

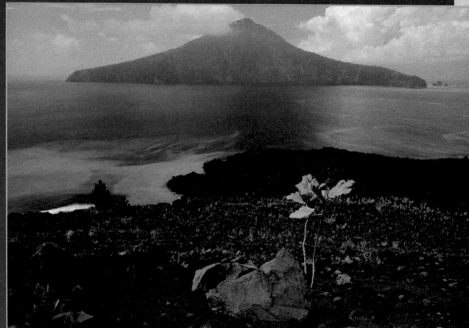

▲ The black sand beach of this Polynesian island is sediment weathered and eroded from the igneous rock of a volcano nearby.

▲ This alluvial fan on the edge of Death Valley, California, was formed when gravel, sand, and finer sediments were deposited by a stream emerging from a mountain canyon.

▲ Layers of shale and chalk form Kansas's Monument Rocks. They are remnants of sediments deposited on the floor of the ancient sea that once covered much of this region.

▲ Heat and pressure deep below Earth's surface can change rock into metamorphic rock, like this banded gneiss.

Matter and the Rock Cycle The rock cycle, illustrated in **Figure 3,** shows how rock can be weathered to small rock and mineral grains. This material then can be eroded and carried away by wind, water, or ice. When you think of erosion, it might seem that the material is somehow destroyed and lost from the cycle. This is not the case. The chemical elements that make up minerals and rocks are not destroyed. This fact illustrates the principle of conservation of matter. The changes that take place in the rock cycle never destroy or create matter. The elements are just redistributed in other forms.

Reading Check *What is the principle of conservation of matter?*

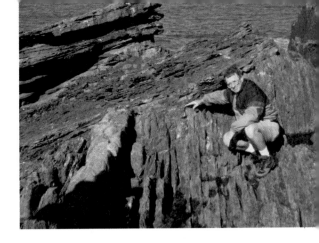

Figure 4 The rock formations at Siccar Point, Scotland, show that rocks undergo constant change.

Discovering the Rock Cycle James Hutton, a Scottish physician and naturalist, first recognized in 1788 that rocks undergo profound changes. Hutton noticed, among other things, that some layers of solid rock in Siccar Point, shown in **Figure 4,** had been altered since they formed. Instead of showing a continuous pattern of horizontal layering, some of the rock layers at Siccar Point are tilted and partly eroded. However, the younger rocks above them are nearly horizontal.

Hutton published these and other observations, which proved that rocks are subject to constant change. Hutton's early recognition of the rock cycle continues to influence geologists.

section 1 review

Summary

What is a rock?
- Rocks are mixtures of minerals, rock fragments, organic matter, volcanic glass, and other materials found in nature.

The Rock Cycle
- The three major types of rock are igneous, metamorphic, and sedimentary.
- Rock cycle processes do not create or destroy matter.
- Processes that are part of the rock cycle change rocks slowly over time.
- In the late eighteenth century, James Hutton recognized some rock cycle processes by observing rocks in the field.
- Some of Hutton's ideas continue to influence geologic thinking today.

Self Check

1. **Explain** how rocks differ from minerals.
2. **Compare and contrast** igneous and metamorphic rock formation.
3. **Describe** the major processes of the rock cycle.
4. **Explain** one way that the rock cycle can illustrate the principle of conservation of matter.
5. **Think Critically** How would you define magma based on the illustration in **Figure 2**? How would you define sediment and sedimentary rock?

Applying Skills

6. **Communicate** Review the model of the rock cycle in **Figure 2.** In your Science Journal, write a story or poem that explains what can happen to a sedimentary rock as it changes throughout the rock cycle.

section 2
Igneous Rocks

as you read

What You'll Learn
- **Recognize** magma and lava as the materials that cool to form igneous rocks.
- **Contrast** the formation of intrusive and extrusive igneous rocks.
- **Contrast** granitic and basaltic igneous rocks.

Why It's Important
Igneous rocks are the most abundant kind of rock in Earth's crust. They contain many valuable resources.

Review Vocabulary
element: substance made of one type of atom that cannot be broken down by ordinary chemical or physical means

New Vocabulary
- igneous rock
- lava
- intrusive
- extrusive
- basaltic
- granitic

Formation of Igneous Rocks

Perhaps you've heard of recent volcanic eruptions in the news. When some volcanoes erupt, they eject a flow of molten rock material, as shown in **Figure 5.** Molten rock material, called magma, flows when it is hot and becomes solid when it cools. When hot magma cools and hardens, it forms **igneous (IHG nee us) rock.** Why do volcanoes erupt, and where does the molten material come from?

Magma In certain places within Earth, the temperature and pressure are just right for rocks to melt and form magma. Most magmas come from deep below Earth's surface. Magma is located at depths ranging from near the surface to about 150 km below the surface. Temperatures of magmas range from about 650°C to 1,200°C, depending on their chemical compositions and pressures exerted on them.

The heat that melts rocks comes from sources within Earth's interior. One source is the decay of radioactive elements within Earth. Some heat is left over from the formation of the planet, which originally was molten. Radioactive decay of elements contained in rocks balances some heat loss as Earth continues to cool.

Because magma is less dense than surrounding solid rock, it is forced upward toward the surface, as shown in **Figure 6.** When magma reaches Earth's surface and flows from volcanoes, it is called **lava.**

Figure 5 Some lava is highly fluid and free-flowing, as shown by this spectacular lava fall in Volcano National Park, East Rift, Kilauea, Hawaii.

Figure 6 Intrusive rocks form from magma trapped below Earth's surface. Extrusive rocks form from lava flowing at the surface.

Intrusive Rocks Magma is melted rock material composed of common elements and fluids. As magma cools, atoms and compounds in the liquid rearrange themselves into new crystals called mineral grains. Rocks form as these mineral grains grow together. Rocks that form from magma below the surface, as illustrated in **Figure 6,** are called **intrusive** igneous rocks. Intrusive rocks are found at the surface only after the layers of rock and soil that once covered them have been removed by erosion. Erosion occurs when the rocks are pushed up by forces within Earth. Because intrusive rocks form at depth and they are surrounded by other rocks, it takes a long time for them to cool. Slowly cooled magma produces individual mineral grains that are large enough to be observed with the unaided eye.

Extrusive Rocks **Extrusive** igneous rocks are formed as lava cools on the surface of Earth. When lava flows on the surface, as illustrated in **Figure 6,** it is exposed to air and water. Lava, such as the basaltic lava shown in **Figure 5,** cools quickly under these conditions. The quick cooling rate keeps mineral grains from growing large, because the atoms in the liquid don't have the time to arrange into large crystals. Therefore, extrusive igneous rocks are fine grained.

 What controls the grain size of an igneous rock?

Table 1 Common Igneous Rocks

Magma Type	Basaltic	Andesitic	Granitic
Intrusive	Gabbro	Diorite	Granite
Extrusive	Basalt, Scoria	Andesite	Rhyolite, Pumice, Obsidian

Volcanic Glass Pumice, obsidian, and scoria are examples of volcanic glass. These rocks cooled so quickly that few or no mineral grains formed. Most of the atoms in these rocks are not arranged in orderly patterns, and few crystals are present.

In the case of pumice and scoria, gases become trapped in the gooey molten material as it cools. Some of these gases eventually escape, but holes are left behind where the rock formed around the pockets of gas.

Classifying Igneous Rocks

Igneous rocks are intrusive or extrusive depending on how they are formed. A way to further classify these rocks is by the magma from which they form. As shown in **Table 1,** an igneous rock can form from basaltic, andesitic, or granitic magma. The type of magma that cools to form an igneous rock determines important chemical and physical properties of that rock. These include mineral composition, density, color, and melting temperature.

Topic: Rock Formation
Visit bookf.msscience.com for Web links to information about intrusive and extrusive rocks.

Activity List several geographic settings where intrusive or extrusive rocks are found. Select one setting for intrusive rocks, and one for extrusive rocks. Describe how igneous rocks form in the two settings, and locate an example of each on a map.

Reading Check *Name two ways igneous rocks are classified.*

Basaltic Rocks Basaltic (buh SAWL tihk) igneous rocks are dense, dark-colored rocks. They form from magma that is rich in iron and magnesium and poor in silica, which is the compound SiO_2. The presence of iron and magnesium in minerals in basalt gives basalt its dark color. Basaltic lava is fluid and flows freely from volcanoes in Hawaii, such as Kilauea. How does this explain the black beach sand common in Hawaii?

Granitic Rocks Granitic igneous rocks are light-colored rocks of a lower density than basaltic rocks. Granitic magma is thick and stiff and contains lots of silica but lesser amounts of iron and magnesium. Because granitic magma is stiff, it can build up a great deal of gas pressure, which is released explosively during violent volcanic eruptions.

Andesitic Rocks Andesitic igneous rocks have mineral compositions between those of basaltic and granitic rocks. Many volcanoes around the rim of the Pacific Ocean formed from andesitic magmas. Like volcanoes that erupt granitic magma, these volcanoes also can erupt violently.

Take another look at **Table 1.** Basalt forms at the surface of Earth because it is an extrusive rock. Granite forms below Earth's surface from magma with a high concentration of silica. When you identify an igneous rock, you can infer how it formed and the type of magma that it formed from.

Melting Rock Inside Earth, materials contained in rocks can melt. In your Science Journal, describe what is happening to the atoms and molecules to cause this change of state.

section 2 review

Summary

Formation of Igneous Rocks
- When molten rock material, called magma, cools and hardens, igneous rock forms.
- Intrusive igneous rocks form as magma cools and hardens slowly, beneath Earth's surface.
- Extrusive igneous rocks form as lava cools and hardens rapidly, at or above Earth's surface.

Classifying Igneous Rocks
- Igneous rocks are further classified according to their mineral compositions.
- The violent nature of some volcanic eruptions is partly explained by the composition of the magma that feeds them.

Self Check

1. **Explain** why some types of magma form igneous rocks that are dark colored and dense.
2. **Identify** the property of magma that causes it to be forced upward toward Earth's surface.
3. **Explain** The texture of obsidian is best described as glassy. Why does obsidian contain few or no mineral grains?
4. **Think Critically** Study the photos in **Table 1.** How are granite and rhyolite similar? How are they different?

Applying Skills

5. **Make and Use Graphs** Four elements make up most of the rocks in Earth's crust. They are: *oxygen—46.6 percent, aluminum—8.1 percent, silicon—27.7 percent,* and *iron—5.0 percent.* Make a bar graph of these data. What might you infer from the low amount of iron?

Igneous Rock Clues

You've learned how color often is used to estimate the composition of an igneous rock. The texture of an igneous rock describes its overall appearance, including mineral grain sizes and the presence or absence of bubble holes, for example. In most cases, grain size relates to how quickly the magma or lava cooled. Crystals you can see without a magnifying lens indicate slower cooling. Smaller, fine-grained crystals indicate quicker cooling, possibly due to volcanic activity. Rocks with glassy textures cooled so quickly that there was no time to form mineral grains.

Real-World Question
What does an igneous rock's texture and color indicate about its formation history?

Goals
- **Classify** different samples of igneous rocks by color and infer their composition.
- **Observe** the textures of igneous rocks and infer how they formed.

Materials
rhyolite granite
basalt obsidian
vesicular basalt gabbro
pumice magnifying lens

Safety Precautions

WARNING: *Some rock samples might have sharp edges. Always use caution while handling samples.*

Procedure
1. **Arrange** rocks according to color (light or dark). Record your observations in your Science Journal.
2. **Arrange** rocks according to similar texture. Consider grain sizes and shapes, presence of holes, etc. Use your magnifying lens to see small features more clearly. Record your observations.

Conclude and Apply
1. **Infer** which rocks are granitic based on color.
2. **Infer** which rocks cooled quickly. What observations led you to this inference?
3. **Identify** any samples that suggest gases were escaping from them as they cooled.
4. **Describe** Which samples have a glassy appearance? How did these rocks form?
5. **Infer** which samples are not volcanic. Explain.

Communicating Your Data
Research the compositions of each of your samples. Did the colors of any samples lead you to infer the wrong compositions? Communicate to your class what you learned.

section 3
Metamorphic Rocks

Formation of Metamorphic Rocks

Have you ever packed your lunch in the morning and not been able to recognize it at lunchtime? You might have packed a sandwich, banana, and a large bottle of water. You know you didn't smash your lunch on the way to school. However, you didn't think about how the heavy water bottle would damage your food if the bottle was allowed to rest on the food all day. The heat in your locker and the pressure from the heavy water bottle changed your sandwich. Like your lunch, rocks can be affected by changes in temperature and pressure.

Metamorphic Rocks Rocks that have changed because of changes in temperature and pressure or the presence of hot, watery fluids are called **metamorphic rocks.** Changes that occur can be in the form of the rock, shown in **Figure 7,** the composition of the rock, or both. Metamorphic rocks can form from igneous, sedimentary, or other metamorphic rocks. What Earth processes can change these rocks?

as you read

What You'll Learn
- **Describe** the conditions in Earth that cause metamorphic rocks to form.
- **Classify** metamorphic rocks as foliated or nonfoliated.

Why It's Important
Metamorphic rocks are useful because of their unique properties.

Review Vocabulary
pressure: the amount of force exerted per unit of area

New Vocabulary
- metamorphic rock
- foliated
- nonfoliated

+ pressure →

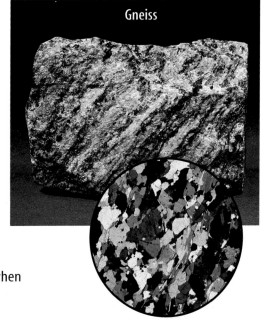

Figure 7 The mineral grains in granite are flattened and aligned when heat and pressure are applied to them. As a result, gneiss is formed.
Describe *other conditions that can cause metamorphic rocks to form.*

Topic: Shale Metamorphism

Visit bookf.msscience.com for Web links to information about the metamorphism of shale. Communicate to your class what you learn.

Activity Make a table with headings that are major rock types that form from shale metamorphism. Under each rock heading, make a list of minerals that can occur in the rock.

Heat and Pressure Rocks beneath Earth's surface are under great pressure from rock layers above them. Temperature also increases with depth in Earth. In some places, the heat and pressure are just right to cause rocks to melt and magma to form. In other areas where melting doesn't occur, some mineral grains can change by dissolving and recrystallizing—especially in the presence of fluids. Sometimes, under these conditions, minerals exchange atoms with surrounding minerals and new, bigger minerals form.

Depending upon the amount of pressure and temperature applied, one type of rock can change into several different metamorphic rocks, and each type of metamorphic rock can come from several kinds of parent rocks. For example, the sedimentary rock shale will change into slate. As increasing pressure and temperature are applied, the slate can change into phyllite, then schist, and eventually gneiss. Schist also can form when basalt is metamorphosed, or changed, and gneiss can come from granite.

 How can one type of rock change into several different metamorphic rocks?

Hot Fluids Did you know that fluids can move through rock? These fluids, which are mostly water with dissolved elements and compounds, can react chemically with a rock and change its composition, especially when the fluids are hot. That's what happens when rock surrounding a hot magma body reacts with hot fluids from the magma, as shown in **Figure 8.** Most fluids that transform rocks during metamorphic processes are hot and mainly are comprised of water and carbon dioxide.

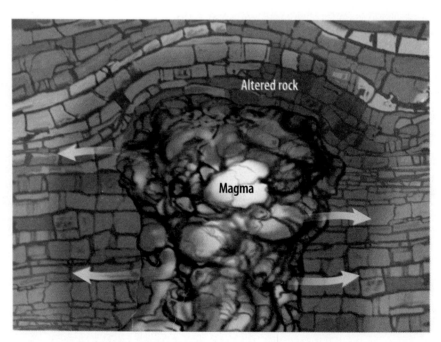

Figure 8 In the presence of hot, water-rich fluids, solid rock can change in mineral composition without having to melt.

Classifying Metamorphic Rocks

Metamorphic rocks form from igneous, sedimentary, or other metamorphic rocks. Heat, pressure, and hot fluids trigger the changes. Each resulting rock can be classified according to its composition and texture.

Foliated Rocks When mineral grains line up in parallel layers, the metamorphic rock is said to have a **foliated** texture. Two examples of foliated rocks are slate and gneiss. Slate forms from the sedimentary rock shale. The minerals in shale arrange into layers when they are exposed to heat and pressure. As **Figure 9** shows, slate separates easily along these foliation layers.

The minerals in slate are pressed together so tightly that water can't pass between them easily. Because it's watertight, slate is ideal for paving around pools and patios. The naturally flat nature of slate and the fact that it splits easily make it useful for roofing and tiling many surfaces.

Gneiss (NISE), another foliated rock, forms when granite and other rocks are changed. Foliation in gneiss shows up as alternating light and dark bands. Movement of atoms has separated the dark minerals, such as biotite mica, from the light minerals, which are mainly quartz and feldspar.

Figure 9 Slate often is used as a building or landscaping material. **Identify** *the properties that make slate so useful for these purposes.*

 What type of metamorphic rock is composed of mineral grains arranged in parallel layers?

Figure 10 This exhibit in Vermont shows the beauty of carved marble.

Nonfoliated Rocks In some metamorphic rocks, layering does not occur. The mineral grains grow and rearrange, but they don't form layers. This process produces a **nonfoliated** texture.

Sandstone is a sedimentary rock that's often composed mostly of quartz grains. When sandstone is heated under a lot of pressure, the grains of quartz grow in size and become interlocking, like the pieces of a jigsaw puzzle. The resulting rock is called quartzite.

Marble is another nonfoliated metamorphic rock. Marble forms from the sedimentary rock limestone, which is composed of the mineral calcite. Usually, marble contains several other minerals besides calcite. For example, hornblende and serpentine give marble a black or greenish tone, whereas hematite makes it red. As **Figure 10** shows, marble is a popular material for artists to sculpt because it is not as hard as other rocks.

So far, you've investigated only a portion of the rock cycle. You still haven't observed how sedimentary rocks are formed and how igneous and metamorphic rocks evolve from them. The next section will complete your investigation of the rock cycle.

section 3 review

Summary

Formation of Metamorphic Rocks

- Changes in pressure, temperature, or the presence of fluids can cause metamorphic rocks to form.
- Rock, altered by metamorphic processes at high temperatures and pressures, changes in the solid state without melting.
- Hot fluids that move through and react with preexisting rock are composed mainly of water and carbon dioxide.
- One source of hot, watery fluids is magma bodies close to the changing rock.
- Any parent rock type—igneous, metamorphic, or sedimentary—can become a metamorphic rock.

Classifying Metamorphic Rocks

- Texture and mineral composition determine how a metamorphic rock is classified.
- Physical properties of metamorphic rocks, such as the watertight nature of slate, make them useful for many purposes.

Self Check

1. **Explain** what role fluids play in rock metamorphism.
2. **Describe** how metamorphic rocks are classified. What are the characteristics of rocks in each of these classifications?
3. **Identify** Give an example of a foliated and a nonfoliated metamorphic rock. Name one of their possible parent rocks.
4. **Think Critically** Marble is a common material used to make sculptures, but not just because it's a beautiful stone. What properties of marble make it useful for this purpose?

Applying Skills

5. **Concept Map** Put the following events in an events-chain concept map that explains how a metamorphic rock might form from an igneous rock. *Hint: Start with "Igneous Rock Forms."* Use each event just once.

   Events: *sedimentary rock forms, weathering occurs, heat and pressure are applied, igneous rock forms, metamorphic rock forms, erosion occurs, sediments are formed, deposition occurs*

section 4
Sedimentary Rocks

Formation of Sedimentary Rocks

Igneous rocks are the most common rocks on Earth, but because most of them exist below the surface, you might not have seen too many of them. That's because 75 percent of the rocks exposed at the surface are sedimentary rocks.

Sediments are loose materials such as rock fragments, mineral grains, and bits of shell that have been moved by wind, water, ice, or gravity. If you look at the model of the rock cycle, you will see that sediments come from already-existing rocks that are weathered and eroded. **Sedimentary rock** forms when sediments are pressed and cemented together, or when minerals form from solutions.

Stacked Rocks Sedimentary rocks often form as layers. The older layers are on the bottom because they were deposited first. Sedimentary rock layers are a lot like the books and papers in your locker. Last week's homework is on the bottom, and today's notes will be deposited on top of the stack. However, if you disturb the stack, the order in which the books and papers are stacked will change, as shown in **Figure 11**. Sometimes, forces within Earth overturn layers of rock, and the oldest are no longer on the bottom.

as you read

What **You'll Learn**
- **Explain** how sedimentary rocks form from sediments.
- **Classify** sedimentary rocks as detrital, chemical, or organic in origin.
- **Summarize** the rock cycle.

Why **It's Important**

Some sedimentary rocks, like coal, are important sources of energy.

Review Vocabulary
weathering: surface processes that work to break down rock mechanically or chemically

New Vocabulary
- sediment
- sedimentary rock
- compaction
- cementation

Figure 11 Like sedimentary rock layers, the oldest paper is at the bottom of the stack. If the stack is disturbed, then it is no longer in order.

Mini LAB

Classifying Sediments

Procedure

WARNING: *Use care when handling sharp objects.*

1. Collect different samples of **sediment**.
2. Spread them on a sheet of **paper**.
3. Use **Table 2** to determine the size range of gravel-sized sediment.
4. Use **tweezers or a dissecting probe** and a **magnifying lens** to separate the gravel-sized sediments.
5. Separate the gravel into piles—rounded or angular.

Analysis

1. Describe the grains in both piles.
2. Determine what rock could form from each type of sediment you have.

Figure 12 During compaction, pore space between sediments decreases, causing them to become packed together more tightly.

Classifying Sedimentary Rocks

Sedimentary rocks can be made of just about any material found in nature. Sediments come from weathered and eroded igneous, metamorphic, and sedimentary rocks. Sediments also come from the remains of some organisms. The composition of a sedimentary rock depends upon the composition of the sediments from which it formed.

Like igneous and metamorphic rocks, sedimentary rocks are classified by their composition and by the manner in which they formed. Sedimentary rocks usually are classified as detrital, chemical, or organic.

Detrital Sedimentary Rocks

The word *detrital* (dih TRI tul) comes from the Latin word *detritus,* which means "to wear away." Detrital sedimentary rocks, such as those shown in **Table 2,** are made from the broken fragments of other rocks. These loose sediments are compacted and cemented together to form solid rock.

Weathering and Erosion When rock is exposed to air, water, or ice, it is unstable and breaks down chemically and mechanically. This process, which breaks rocks into smaller pieces, is called weathering. **Table 2** shows how these pieces are classified by size. The movement of weathered material is called erosion.

Compaction Erosion moves sediments to a new location, where they then are deposited. Here, layer upon layer of sediment builds up. Pressure from the upper layers pushes down on the lower layers. If the sediments are small, they can stick together and form solid rock. This process, shown in **Figure 12,** is called **compaction.**

Reading Check *How do rocks form through compaction?*

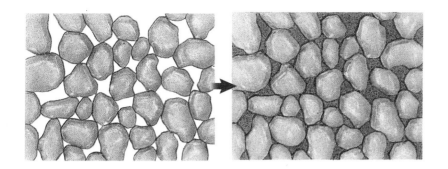

Figure 13 Sediments are cemented together as minerals crystallize between grains.

Cementation If sediments are large, like sand and pebbles, pressure alone can't make them stick together. Large sediments have to be cemented together. As water moves through soil and rock, it picks up materials released from minerals during weathering. The resulting solution of water and dissolved materials moves through open spaces between sediments. **Cementation,** which is shown in **Figure 13,** occurs when minerals such as quartz, calcite, and hematite are deposited between the pieces of sediment. These minerals, acting as natural cements, hold the sediment together like glue, making a detrital sedimentary rock.

Shape and Size of Sediments Detrital rocks have granular textures, much like granulated sugar. They are named according to the shapes and sizes of the sediments that form them. For example, conglomerate and breccia both form from large sediments, as shown in **Table 2.** If the sediments are rounded, the rock is called conglomerate. If the sediments have sharp angles, the rock is called breccia. The roundness of sediment particles depends on how far they have been moved by wind or water.

Table 2 Sediment Sizes and Detrital Rocks				
Sediment	Clay	Silt	Sand	Gravel
Size Range	<0.004 mm	0.004–0.063 mm	0.063–2 mm	>2 mm
Example	Shale	Siltstone	Sandstone	Conglomerate (shown) or Breccia

SECTION 4 Sedimentary Rocks

Figure 14 Although concrete strongly resembles conglomerate, concrete is not a rock because it does not occur in nature.

Materials Found in Sedimentary Rocks The gravel-sized sediments in conglomerate and breccia can consist of any type of rock or mineral. Often, they are composed of chunks of the minerals quartz and feldspar. They also can be pieces of rocks such as gneiss, granite, or limestone. The cement that holds the sediments together usually is made of quartz or calcite.

Have you ever looked at the concrete in sidewalks, driveways, and stepping stones? The concrete in **Figure 14** is made of gravel and sand grains that have been cemented together. Although the structure is similar to that of naturally occurring conglomerate, it cannot be considered a rock.

Sandstone is formed from smaller particles than conglomerates and breccias. Its sand-sized sediments can be just about any mineral, but they are usually grains of minerals such as quartz and feldspar that are resistant to weathering. Siltstone is similar to sandstone except it is made of smaller, silt-sized particles. Shale is a detrital sedimentary rock that is made mainly of clay-sized particles. Clay-sized sediments are compacted together by pressure from overlying layers.

Sedimentary Petrology Research the work done by sedimentary petrologists. Include examples of careers in academia and in industry.

Chemical Sedimentary Rocks

Chemical sedimentary rocks form when dissolved minerals come out of solution. You can show that salt is deposited in the bottom of a glass or pan when saltwater solution evaporates. In a similar way, minerals collect when seas or lakes evaporate. The deposits of minerals that come out of solution form sediments and rocks. For example, the sediment making up New Mexico's White Sands desert consists of pieces of a chemical sedimentary rock called rock gypsum. Chemical sedimentary rocks are different. They are not made from pieces of preexisting rocks.

Reading Check *How do chemical sedimentary rocks form?*

Limestone Calcium carbonate is carried in solution in ocean water. When calcium carbonate ($CaCO_3$) comes out of solution as calcite and its many crystals grow together, limestone forms. Limestone also can contain other minerals and sediments, but it must be at least 50 percent calcite. Limestone usually is deposited on the bottom of lakes or shallow seas. Large areas of the central United States have limestone bedrock because seas covered much of the country for millions of years. It is hard to imagine Kansas being covered by ocean water, but it has happened several times throughout geological history.

Rock Salt When water that is rich in dissolved salt evaporates, it often deposits the mineral halite. Halite forms rock salt, shown in **Figure 15**. Rock salt deposits can range in thickness from a few meters to more than 400 m. Companies mine these deposits because rock salt is an important resource. It's used in the manufacturing of glass, paper, soap, and dairy products. The halite in rock salt is processed and used as table salt.

Figure 15 Rock salt is extracted from this mine in Germany. The same salt can be processed and used to season your favorite foods.

Organic Sedimentary Rocks

Rocks made of the remains of once-living things are called organic sedimentary rocks. One of the most common organic sedimentary rocks is fossil-rich limestone. Like chemical limestone, fossil-rich limestone is made of the mineral calcite. However, fossil-rich limestone mostly contains remains of once-living ocean organisms instead of only calcite that formed directly from ocean water.

Animals such as mussels, clams, corals, and snails make their shells from $CaCO_3$ that eventually becomes calcite. When they die, their shells accumulate on the ocean floor. When these shells are cemented together, fossil-rich limestone forms. If a rock is made completely of shell fragments that you can see, the rock is called coquina (koh KEE nuh).

Chalk Chalk is another organic sedimentary rock that is made of microscopic shells. When you write with naturally occurring chalk, you're crushing and smearing the calcite-shell remains of once-living ocean organisms.

Coal Another useful organic sedimentary rock is coal, shown in **Figure 16.** Coal forms when pieces of dead plants are buried under other sediments in swamps. These plant materials are chemically changed by microorganisms. The resulting sediments are compacted over millions of years to form coal, an important source of energy. Much of the coal in North America and Europe formed during a period of geologic time that is so named because of this important reason. The Carboniferous Period, which spans from approximately 360 to 286 million years ago, was named in Europe. So much coal formed during this interval of time that coal's composition—primarily carbon—was the basis for naming a geologic period.

Applying Math — Calculate Thickness

COAL FORMATION It took 300 million years for a layer of plant matter about 0.9 m thick to produce a bed of bituminous coal 0.3 m thick. Estimate the thickness of plant matter that produced a bed of coal 0.15 m thick.

Solution

1 *This is what you know:*
- original thickness of plant matter = 0.9 m
- original coal thickness = 0.3 m
- new coal thickness = 0.15 m

2 *This is what you need to know:* thickness of plant matter needed to form 0.15 m of coal

3 *This is the equation you need to use:* (thickness of plant matter)/(new coal thickness) = (original thickness of plant matter)/(original coal thickness)

4 *Substitute the known values:* (? m plant matter)/(0.15 m coal) = (0.9 m plant matter)/(0.3 m coal)

5 *Solve the equation:* (? m plant matter) = (0.9 m plant matter)(0.15 m coal)/(0.3 m coal) = 0.45 m plant matter

6 *Check your answer:* Multiply your answer by the original coal thickness. Divide by the original plant matter thickness to get the new coal thickness.

Practice Problems

1. Estimate the thickness of plant matter that produced a bed of coal 0.6 m thick.
2. About how much coal would have been produced from a layer of plant matter 0.50 m thick?

For more practice, visit bookf.msscience.com/math_practice

Figure 16 This coal layer in Alaska is easily identified by its jet-black color, as compared with other sedimentary layers.

Another Look at the Rock Cycle

You have seen that the rock cycle has no beginning and no end. Rocks change continually from one form to another. Sediments can become so deeply buried that they eventually become metamorphic or igneous rocks. These reformed rocks later can be uplifted and exposed to the surface—possibly as mountains to be worn away again by erosion.

All of the rocks that you've learned about in this chapter formed through some process within the rock cycle. All of the rocks around you, including those used to build houses and monuments, are part of the rock cycle. Slowly, they are all changing, because the rock cycle is a continuous, dynamic process.

section 4 review

Summary

Formation of Sedimentary Rocks
- Sedimentary rocks form as layers, with older layers near the bottom of an undisturbed stack.

Classifying Sedimentary Rocks
- To classify a sedimentary rock, determine its composition and texture.

Detrital Sedimentary Rocks
- Rock and mineral fragments make up detrital rocks.

Chemical Sedimentary Rocks
- Chemical sedimentary rocks form from solutions of dissolved minerals.

Organic Sedimentary Rocks
- The remains of once-living organisms make up organic sedimentary rocks.

Self Check

1. **Identify** where sediments come from.
2. **Explain** how compaction is important in the formation of coal.
3. **Compare and contrast** detrital and chemical sedimentary rock.
4. **List** chemical sedimentary rocks that are essential to your health or that are used to make life more convenient. How is each used?
5. **Think Critically** Explain how pieces of granite and slate could both be found in the same conglomerate. How would the granite and slate pieces be held together?

Applying Math

6. **Calculate Ratios** Use information in **Table 2** to estimate how many times larger the largest grains of silt and sand are compared to the largest clay grains.

Sedimentary Rocks

Sedimentary rocks are formed by compaction and cementation of sediment. Because sediment is found in all shapes and sizes, do you think these characteristics could be used to classify detrital sedimentary rocks? Sedimentary rocks also can be classified as chemical or organic.

Goals
- **Observe** sedimentary rock characteristics.
- **Compare and contrast** sedimentary rock textures.
- **Classify** sedimentary rocks as detrital, chemical, or organic.

Materials
unknown sedimentary rock samples
marking pen
5% hydrochloric acid (HCl) solution
dropper
paper towels
water
magnifying lens
metric ruler

Safety Precautions

WARNING: *HCl is an acid and can cause burns. Wear goggles and a lab apron. Rinse spills with water and wash hands afterward.*

● Real-World Question
How are rock characteristics used to classify sedimentary rocks as detrital, chemical, or organic?

● Procedure
1. Make a Sedimentary Rock Samples chart in your Science Journal similar to the one shown on the next page.
2. **Determine** the sizes of sediments in each sample, using a magnifying lens and a metric ruler. Using **Table 2,** classify any grains of sediment in the rocks as gravel, sand, silt, or clay. In general, the sediment is silt if it is gritty and just barely visible, and clay if it is smooth and if individual grains are not visible.
3. Place a few drops of 5% HCl solution on each rock sample. Bubbling on a rock indicates the presence of calcite.
4. **Examine** each sample for fossils and describe any that are present.
5. **Determine** whether each sample has a granular or nongranular texture.

Using Scientific Methods

Sedimentary Rock Samples

Sample	Observations	Minerals or Fossils Present	Sediment Size	Detrital, Chemical, or Organic	Rock Name
A					
B			Do not write in this book.		
C					
D					
E					

Analyze Your Data

1. **Classify** your samples as detrital, chemical, or organic.
2. **Identify** each rock sample.

Conclude and Apply

1. **Explain** why you tested the rocks with acid. What minerals react with acid?
2. **Compare and contrast** sedimentary rocks that have a granular texture with sedimentary rocks that have a nongranular texture.

Communicating Your Data

Compare your conclusions with those of other students in your class. **For more help, refer to the** Science Skill Handbook.

TIME SCIENCE AND Society

SCIENCE ISSUES THAT AFFECT YOU!

Australia's controversial rock star

One of the most famous rocks in the world is causing serious problems for Australians

Uluru (yew LEW rew), also known as Ayers Rock, is one of the most popular tourist destinations in Australia. This sandstone skyscraper is more than 8 km around, over 300 m high, and extends as much as 4.8 km below the surface. One writer describes it as an iceberg in the desert. Geologists hypothesize that the mighty Uluru rock began forming 550 million years ago during Precambrian time. That's when large mountain ranges started to form in Central Australia.

For more than 25,000 years, this geological wonder has played an important role in the lives of the Aboriginal peoples, the Anangu (a NA noo). These native Australians are the original owners of the rock and have spiritual explanations for its many caves, holes, and scars.

Tourists Take Over

In the 1980s, some 100,000 tourists visited—and many climbed—Uluru. In 2000, the rock attracted about 400,000 tourists. The Anangu take offense at anyone climbing their sacred rock. However, if climbing the rock were outlawed, tourism would be seriously hurt. That would mean less income for Australians.

To respect the Anangu's wishes, the Australian government returned Ayers Rock to the Anangu

Athlete Nova Benis-Kneebone had the honor of receiving the Olympic torch near the sacred Uluru and carried it partway to the Olympic stadium.

in 1985 and agreed to call it by its traditional name. The Anangu leased back the rock to the Australian government until the year 2084, when its management will return to the Anangu. Until then, the Anangu will collect 25 percent of the money people pay to visit the rock.

The Aboriginal people encourage tourists to respect their beliefs. They offer a walking tour around the rock, and they show videos about Aboriginal traditions. The Anangu sell T-shirts that say "I *didn't* climb Uluru." They hope visitors to Uluru will wear the T-shirt with pride and respect.

Write Research a natural landmark or large natural land or water formation in your area. What is the geology behind it? When was it formed? How was it formed? Write a folktale that explains its formation. Share your folktale with the class.

For more information, visit bookf.msscience.com/time

chapter 2 Study Guide

Reviewing Main Ideas

Section 1 The Rock Cycle

1. A rock is a mixture of one or more minerals, rock fragments, organic matter, or volcanic glass.
2. The rock cycle includes all processes by which rocks form.

Section 2 Igneous Rocks

1. Magma and lava are molten materials that harden to form igneous rocks.
2. Intrusive igneous rocks form when magma cools slowly below Earth's surface. Extrusive igneous rocks form when lava cools rapidly at the surface.
3. The compositions of most igneous rocks range from granitic to andesitic to basaltic.

Section 3 Metamorphic Rocks

1. Heat, pressure, and fluids can cause metamorphic rocks to form.
2. Slate and gneiss are examples of foliated metamorphic rocks. Quartzite and marble are examples of nonfoliated metamorphic rocks.

Section 4 Sedimentary Rocks

1. Detrital sedimentary rocks form when fragments of rocks and minerals are compacted and cemented together.
2. Chemical sedimentary rocks come out of solution or are left behind by evaporation.
3. Organic sedimentary rocks contain the remains of once-living organisms.

Visualizing Main Ideas

Copy and complete the following concept map on rocks. Use the following terms: organic, metamorphic, foliated, extrusive, igneous, and chemical.

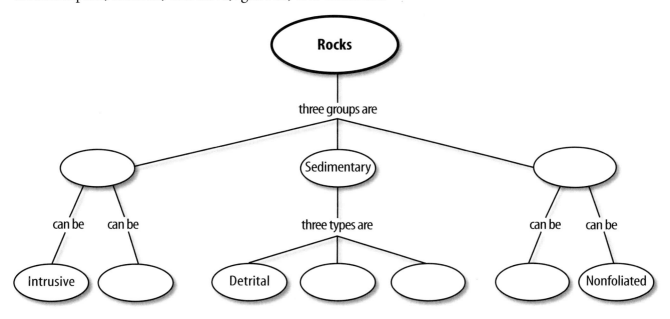

Chapter 2 Review

Using Vocabulary

basaltic p. 43
cementation p. 51
compaction p. 50
extrusive p. 41
foliated p. 47
granitic p. 43
igneous rock p. 40
intrusive p. 41
lava p. 40
metamorphic rock p. 45
nonfoliated p. 48
rock p. 36
rock cycle p. 37
sediment p. 49
sedimentary rock p. 49

Explain the difference between the vocabulary words in each of the following sets.

1. foliated—nonfoliated
2. cementation—compaction
3. sediment—lava
4. extrusive—intrusive
5. rock—rock cycle
6. metamorphic rock—igneous rock—sedimentary rock
7. sediment—sedimentary rock
8. lava—igneous rock
9. rock—sediment
10. basaltic—granitic

Checking Concepts

Choose the word or phrase that best answers the question.

11. Why does magma tend to rise toward Earth's surface?
    A) It is more dense than surrounding rocks.
    B) It is more massive than surrounding rocks.
    C) It is cooler than surrounding rocks.
    D) It is less dense than surrounding rocks.

12. During metamorphism of granite into gneiss, what happens to minerals?
    A) They partly melt.
    B) They become new sediments.
    C) They grow smaller.
    D) They align into layers.

13. Which rock has large mineral grains?
    A) granite   C) obsidian
    B) basalt   D) pumice

14. Which type of rock is shown in this photo?
    A) foliated
    B) nonfoliated
    C) intrusive
    D) extrusive

15. What do igneous rocks form from?
    A) sediments   C) gravel
    B) mud   D) magma

16. What sedimentary rock is made of large, angular pieces of sediments?
    A) conglomerate   C) limestone
    B) breccia   D) chalk

17. Which of the following is an example of a detrital sedimentary rock?
    A) limestone   C) breccia
    B) evaporite   D) chalk

18. What is molten material at Earth's surface called?
    A) limestone   C) breccia
    B) lava   D) granite

19. Which of these is an organic sedimentary rock?
    A) coquina   C) rock salt
    B) sandstone   D) conglomerate

chapter 2 Review

Thinking Critically

20. **Infer** Granite, pumice, and scoria are igneous rocks. Why doesn't granite have airholes like the other two?

21. **Infer** why marble rarely contains fossils.

22. **Predict** Would you expect quartzite or sandstone to break more easily? Explain your answer.

23. **Compare and contrast** basaltic and granitic magmas.

24. **Form Hypotheses** A geologist was studying rocks in a mountain range. She found a layer of sedimentary rock that had formed in the ocean. Hypothesize how this could happen.

25. **Concept Map** Copy and complete the concept map shown below. Use the following terms and phrases: *magma, sediments, igneous rock, sedimentary rock, metamorphic rock.* Add and label any missing arrows.

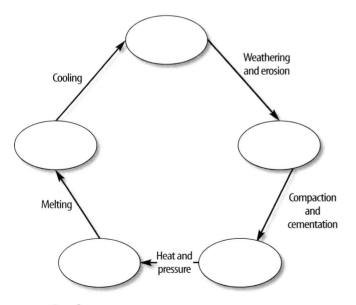

Performance Activities

26. **Poster** Collect a group of rocks. Make a poster that shows the classifications of rocks, and glue your rocks to the poster under the proper headings. Describe your rocks and explain where you found them.

Applying Math

27. **Grain Size** Assume that the conglomerate shown on the second page of the "Sedimentary Rocks" lab is one-half of its actual size. Determine the average length of the gravel in the rock.

28. **Plant Matter** Suppose that a 4-m layer of plant matter was compacted to form a coal layer 1 m thick. By what percent has the thickness of organic material been reduced?

Use the graph below to answer questions 29 and 30.

29. **Melting Granite** Determine the melting temperature of a water-rich granite at a pressure of 0.2 GPa.

    Pressure conversions:
    1 GPa, or gigapascal, = 10,000 bars
    1 bar = 0.9869 atmospheres

30. **Melting Pressure** At about what pressure will a water-rich granite melt at 680°C?

chapter 2 Standardized Test Practice

Part 1 | Multiple Choice

Record your answers on the answer sheet provided by your teacher or on a sheet of paper.

Use the illustration below to answer question 1.

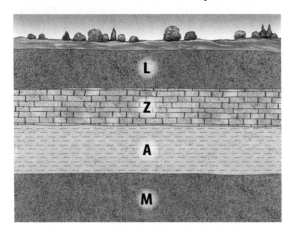

1. These layers of sedimentary rock were not disturbed after they were deposited. Which layer was deposited first?
   A. layer L   C. layer M
   B. layer Z   D. layer A

2. Who realized that rocks undergo changes through long periods of time after observing rocks at Siccar Point, Scotland?
   A. James Hutton   C. Galileo Galilei
   B. Neil Armstrong D. Albert Einstein

3. During which process do minerals precipitate in the spaces between sediment grains?
   A. compaction   C. cementation
   B. weathering   D. conglomerate

4. Which rock often is sculpted to create statues?
   A. shale   C. coquina
   B. marble   D. conglomerate

Test-Taking Tip

Careful Reading Read each question carefully for full understanding.

5. Which of the following rocks is a metamorphic rock?
   A. shale   C. slate
   B. granite   D. pumice

6. Which rock consists mostly of pieces of seashell?
   A. sandstone   C. pumice
   B. coquina   D. granite

Use the diagram below to answer questions 7–9.

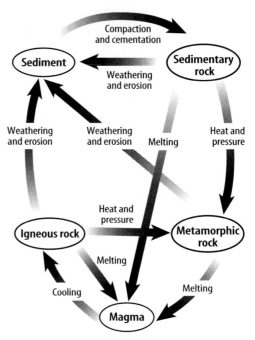

7. Which process in the rock cycle causes magma to form?
   A. melting   C. weathering
   B. erosion   D. cooling

8. What forms when rocks are weathered and eroded?
   A. igneous rock   C. sedimentary rock
   B. sediment   D. metamorphic rock

9. Which type of rock forms because of high heat and pressure without melting?
   A. igneous rock   C. sedimentary rock
   B. intrusive rock D. metamorphic rock

Standardized Test Practice

Part 2 Short Response/Grid In

Record your answers on the answer sheet provided by your teacher or on a sheet of paper.

10. What is a rock? How is a rock different from a mineral?
11. Explain why some igneous rocks are coarse and others are fine.
12. What is foliation? How does it form?
13. How do chemical sedimentary rocks, such as rock salt, form?
14. Why do some rocks contain fossils?
15. How is the formation of chemical sedimentary rocks similar to the formation of cement in detrital sedimentary rocks?

Use the graph below to answer questions 16–17.

Temperature Beneath the Continents

16. According to the graph, about how deep below a continent does the temperature reach 1,000°C?
17. In general, what happens to temperature as depth below Earth's surface increases?

Part 3 Open Ended

Record your answers on a sheet of paper.

Use the table below to answer questions 18 and 19.

Magma Type	Basaltic	Andesitic	Granitic
Intrusive			
Extrusive			

Do not write in this book.

18. Copy the table on your paper. Then, fill in the empty squares with a correct rock name.
19. Explain how igneous rocks are classified.
20. Explain how loose sediment can become sedimentary rock.
21. Why does pressure increase with depth in Earth? How does higher pressure affect rocks?
22. Why is slate sometimes used as shingles for roofs? What other rocks are used for important purposes in society?
23. How are organic sedimentary rocks different from other rocks? List an example of an organic sedimentary rock.
24. Why is the rock cycle called a cycle?
25. A geologist found a sequence of rocks in which 200-million-year-old shales were on top of 100-million-year-old sandstones. Hypothesize how this could happen.
26. Explain why coquina could be classified in more than one way.

chapter 3

Earth's Energy and Mineral Resources

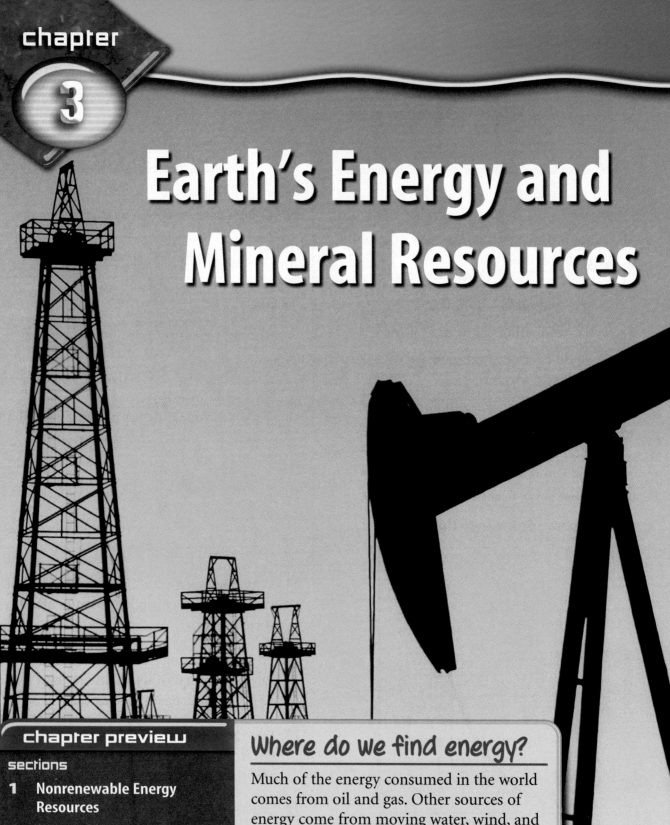

chapter preview

sections
1. **Nonrenewable Energy Resources**
2. **Renewable Energy Resources**
   Lab Soaking Up Solar Energy
3. **Mineral Resources**
   Lab Home Sweet Home
   Virtual Lab What are the advantages of alternative energy sources?

Where do we find energy?
Much of the energy consumed in the world comes from oil and gas. Other sources of energy come from moving water, wind, and the Sun's rays. In this chapter you'll learn about many types of energy resources and the importance of conserving these resources.

Science Journal Write three ways electricity is generated at a power plant.

Start-Up Activities

Finding Energy Reserves

The physical properties of Earth materials determine how easily liquids and gases move through them. Geologists use these properties, in part, to predict where reserves of energy resources like petroleum or natural gas can be found.

1. Obtain a sample of sandstone and a sample of shale from your teacher.
2. Make sure that your samples can be placed on a tabletop so that the sides facing up are reasonably flat and horizontal.
3. Place the two samples side by side in a shallow baking pan.
4. Using a dropper, place three drops of cooking oil on each sample.
5. For ten minutes, observe what happens to the oil on the samples.
6. **Think Critically** Write your observations in your Science Journal. Infer which rock type might be a good reservoir for petroleum.

Preview this chapter's content and activities at
bookf.msscience.com

Energy Resources Make the following Foldable to help you identify energy resources.

STEP 1 Fold a sheet of paper in half lengthwise. Make the back edge about 1.25 cm longer than the front edge.

STEP 2 Turn lengthwise and fold into thirds.

STEP 3 Unfold and cut only the top layer along both folds to make three tabs.

STEP 4 Label each tab as shown.

Find Main Ideas As you read the chapter, list examples on the front of the tabs and write about each type of resource under the tabs.

section 1
Nonrenewable Energy Resources

as you read

What You'll Learn
- **Identify** examples of nonrenewable energy resources.
- **Describe** the advantages and disadvantages of using fossil fuels.
- **Explain** the advantages and disadvantages of using nuclear energy.

Why It's Important
Nonrenewable resources should be conserved to ensure their presence for future generations.

Review Vocabulary
fuel: a material that provides useful energy

New Vocabulary
- fossil fuel
- coal
- oil
- natural gas
- reserve
- nuclear energy

Energy

The world's population relies on energy of all kinds. Energy is the ability to cause change. Some energy resources on Earth are being used faster than natural Earth processes can replace them. These resources are referred to as nonrenewable energy resources. Most of the energy resources used to generate electricity are nonrenewable.

Fossil Fuels

Nonrenewable energy resources include fossil fuels. **Fossil fuels** are fuels such as coal, oil, and natural gas that form from the remains of plants and other organisms that were buried and altered over millions of years. Coal is a sedimentary rock formed from the compacted and transformed remains of ancient plant matter. Oil is a liquid hydrocarbon that often is referred to as petroleum. Hydrocarbons are compounds that contain hydrogen and carbon atoms. Other naturally occurring hydrocarbons occur in the gas or semisolid states. Fossil fuels are processed to make gasoline for cars, to heat homes, and for many other uses, as shown in **Table 1**.

Table 1 Uses of Fossil Fuels	
Coal	■ To generate electricity
Oil	■ To produce gasoline and other fuels ■ As lubricants ■ To make plastics, home shingles, and other products
Natural Gas	■ To heat buildings ■ As a source of sulfur

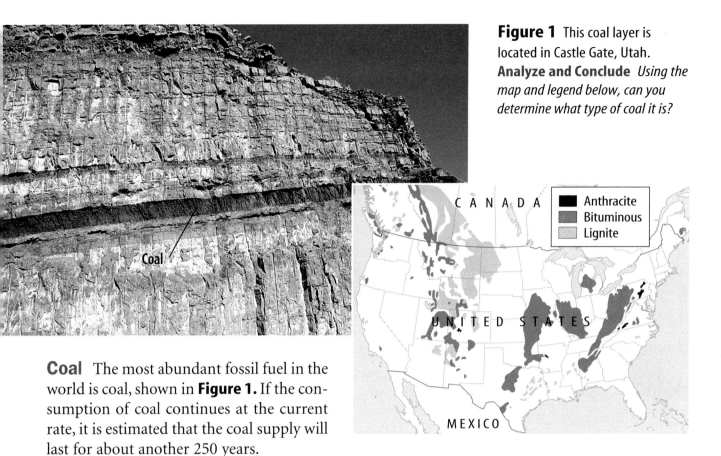

Figure 1 This coal layer is located in Castle Gate, Utah. **Analyze and Conclude** Using the map and legend below, can you determine what type of coal it is?

Coal The most abundant fossil fuel in the world is coal, shown in **Figure 1.** If the consumption of coal continues at the current rate, it is estimated that the coal supply will last for about another 250 years.

Coal is a rock that contains at least 50 percent plant remains. Coal begins to form when plants die in a swampy area. The dead plants are covered by more plants, water, and sediment, preventing atmospheric oxygen from coming into contact with the plant matter. The lack of atmospheric oxygen prevents the plant matter from decaying rapidly. Bacterial growth within the plant material causes a gradual breakdown of molecules in the plant tissue, leaving carbon and some impurities behind. This is the material that eventually will become coal after millions of years. Bacteria also cause the release of methane gas, carbon dioxide, ammonia, and water as the original plant matter breaks down.

 What happens to begin the formation of coal in a swampy area?

Synthetic Fuels Unlike gasoline, which is refined from petroleum, other fuels called synthetic fuels are extracted from solid organic material. Synthetic fuels can be created from coal—a sedimentary rock containing hydrocarbons. The hydrocarbons are extracted from coal to form liquid and gaseous synthetic fuels. Liquid synthetic fuels can be processed to produce gasoline for automobiles and fuel oil for home heating. Gaseous synthetic fuels are used to generate electricity and heat buildings.

Coal Formation The coal found in the eastern and midwestern United States formed from plants that lived in great swamps about 300 million years ago during the Pennsylvanian Period of geologic time. Research the Pennsylvanian Period to find out what types of plants lived in these swamps. Describe the plants in your Science Journal.

SECTION 1 Nonrenewable Energy Resources **F ◆ 67**

Stages of Coal Formation As decaying plant material loses gas and moisture, the concentration of carbon increases. The first step in this process, shown in Figure 2, results in the formation of peat. Peat is a layer of organic sediment. When peat burns, it releases large amounts of smoke because it has a high concentration of water and impurities.

As peat is buried under more sediment, it changes into lignite, which is a soft, brown coal with much less moisture. Heat and pressure produced by burial force water out of peat and concentrate carbon in the lignite. Lignite releases more energy and less smoke than peat when it is burned.

As the layers are buried deeper, bituminous coal, or soft coal, forms. Bituminous coal is compact, black, and brittle. It provides lots of heat energy when burned. Bituminous coal contains various levels of sulfur, which can pollute the environment.

If enough heat and pressure are applied to buried layers of bituminous coal, anthracite coal forms. Anthracite coal contains the highest amount of carbon of all forms of coal. Therefore, anthracite coal is the cleanest burning of all coals.

Figure 2 Coal is formed in four basic stages.

A Dead plant material accumulates in swamps and eventually forms a layer of peat.

B Over time, heat and pressure cause the peat to change into lignite coal.

C As the lignite coal becomes buried by more sediments, heat and pressure change it into bituminous coal.

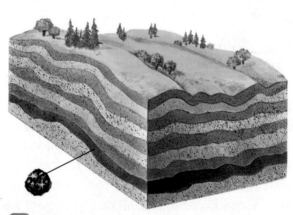

D When bituminous coal is heated and squeezed during metamorphism, anthracite coal forms.

Oil and Natural Gas

Coal isn't the only fossil fuel used to obtain energy. Two other fossil fuels that provide large quantities of the energy used today are oil and natural gas. **Oil** is a thick, black liquid formed from the buried remains of microscopic marine organisms. **Natural gas** forms under similar conditions and often with oil, but it forms in a gaseous state. Oil and natural gas are hydrocarbons. However, natural gas is composed of hydrocarbon molecules that are lighter than those in oil.

Residents of the United States burn vast quantities of oil and natural gas for daily energy requirements. As shown in **Figure 3,** Americans obtain most of their energy from these sources. Natural gas is used mostly for heating and cooking. Oil is used in many ways, including as heating oil, gasoline, lubricants, and in the manufacture of plastics and other important compounds.

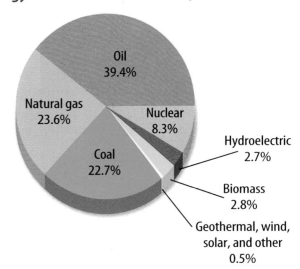

Figure 3 This circle graph shows the percentages of energy that the United States derives from various energy resources. **Calculate** What percentage is from nonrenewable energy resources?

Formation of Oil and Natural Gas

Most geologists agree that petroleum forms over millions of years from the remains of tiny marine organisms in ocean sediment. The process begins when marine organisms called plankton die and fall to the seafloor. Similar to the way that coal is buried, sediment is deposited over them. The temperature rises with depth in Earth, and increased heat eventually causes the dead plankton to change to oil and gas after they have been buried deeply by sediment.

Oil and natural gas often are found in layers of rock that have become tilted or folded. Because they are less dense than water, oil and natural gas are forced upward. Rock layers that are impermeable, such as shale, stop this upward movement. When this happens, a folded shale layer can trap the oil and natural gas below it. Such a trap for oil and gas is shown in **Figure 4.** The rock layer beneath the shale in which the petroleum and natural gas accumulate is called a reservoir rock.

Figure 4 Oil and natural gas are fossil fuels formed by the burial of marine organisms. These fuels can be trapped and accumulate beneath Earth's surface.

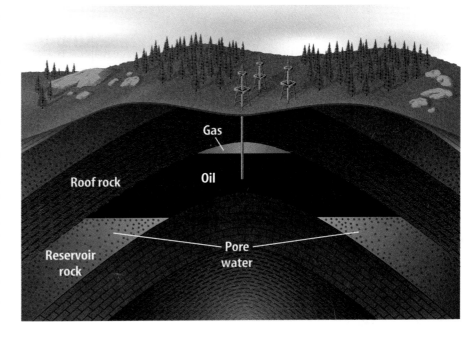

SECTION 1 Nonrenewable Energy Resources **F ◆ 69**

Removing Fossil Fuels from the Ground

Coal is removed from the ground using one of several methods of excavation. The two most common methods are strip mining, also called open-pit mining, and underground mining, shown in **Figure 5**. Oil and natural gas are removed by pumping them out of the ground.

Coal Mining During strip mining, as shown in **Figure 5**, layers of soil and rock above coal are removed and piled to one side. The exposed coal then is removed and loaded into trucks or trains and transported elsewhere. After the coal has been removed, mining companies often return the soil and rock to the open pit and cover it with topsoil. Trees and grass are planted in a process called land reclamation. If possible, animals native to the area are reintroduced. Strip mining is used only when the coal deposits are close to the surface.

In one method of underground coal mining, tunnels are dug and pillars of rock are left to support the rocks surrounding the tunnels. Two types of underground coal mines are drift mines and slope mines. Drift mining, shown in the **Figure 5** inset photo, is the removal of coal that is not close to Earth's surface through a horizontal opening in the side of a hill or mountain. In slope mining, an angled opening and air shaft are made in the side of a mountain to remove coal.

Figure 5 Coal is a fossil fuel that can be removed from Earth in many different ways.

During strip mining, coal is accessed by removing the soil and rock above it.

During drift mining, tunnels are made into Earth.
Explain *how you think the coal is removed from these tunnels.*

Drilling for Oil and Gas Oil and natural gas are fossil fuels that can be pumped from underground deposits. Geologists and engineers drill wells through rocks where these resources might be trapped, as shown in **Figure 6.** As the well is being drilled, it is lined with pipe to prevent it from caving in. When the drill bit reaches the rock layer containing oil, drilling is stopped. Equipment is installed to control the flow of oil. The surrounding rock then is fractured to allow oil and gas to flow into the well. The oil and gas are pumped to the surface.

 How are oil and natural gas brought to Earth's surface?

Fossil Fuel Reserves

The amount of a fossil fuel that can be extracted at a profit using current technology is known as a **reserve.** This is not the same as a fossil fuel resource. A fossil fuel resource has fossil fuels that are concentrated enough that they can be extracted from Earth in useful amounts. However, a resource is not classified as a reserve unless the fuel can be extracted economically. What might cause a known fossil fuel resource to become classified as a reserve?

Methane Hydrates You have learned that current reserves of coal will last about 250 years. Enough natural gas is located in the United States to last about 60 more years. However, recent studies indicate that a new source of methane, which is the main component of natural gas, might be located beneath the seafloor. Icelike substances known as methane hydrates could provide tremendous reserves of methane.

Methane hydrates are stable molecules found hundreds of meters below sea level in ocean floor sediment. They form under conditions of relatively low temperatures and high pressures. The hydrocarbons are trapped within the cagelike structure of ice, as described in **Figure 7.** Scientists estimate that more carbon is contained in methane hydrates than in all current fossil fuel deposits combined. Large accumulations of methane hydrates are estimated to exist off the eastern coast of the United States. Can you imagine what it would mean to the world's energy supply if relatively clean-burning methane could be extracted economically from methane hydrates?

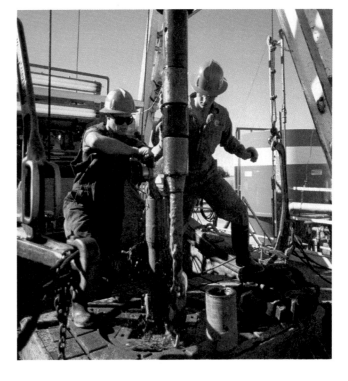

Figure 6 Oil and natural gas are recovered from Earth by drilling deep wells.

Topic: Methane Hydrates
Visit bookf.msscience.com for Web links to information about methane hydrates.

Activity Identify which oceans might contain significant amounts of methane hydrates.

NATIONAL GEOGRAPHIC VISUALIZING METHANE HYDRATES

Figure 7

Reserves of fossil fuels—such as oil, coal, and natural gas—are limited and will one day be used up. Methane hydrates could be an alternative energy source. This icelike substance, background, has been discovered in ocean floor sediments and in permafrost regions worldwide. If scientists can harness this energy, the world's gas supply could be met for years to come.

Methane hydrates are highly flammable compounds made up of methane—the main component of natural gas—trapped in a cage of frozen water. Methane hydrates represent an enormous source of potential energy. However, they contain a greenhouse gas that might intensify global warming. More research is needed to determine how to safely extract them from the seafloor.

In the photo above, a Russian submersible explores a site in the North Atlantic that contains methane hydrate deposits.

Conserving Fossil Fuels Do you sometimes forget to turn off the lights when you walk out of a room? Wasteful habits might mean that electricity to run homes and industries will not always be as plentiful and cheap as it is today. Fossil fuels take millions of years to form and are used much faster than Earth processes can replenish them.

Today, coal provides about 25 percent of the energy that is used worldwide and 22 percent of the energy used in the United States. Oil and natural gas provide almost 61 percent of the world's energy and about 65 percent of the U.S. energy supply. At the rate these fuels are being used, they could run out someday. How can this be avoided?

By remembering to turn off lights and appliances, you can avoid wasting fossil fuels. Another way to conserve fossil fuels is to make sure doors and windows are shut tightly during cold weather so heat doesn't leak out of your home. If you have air-conditioning, run it as little as possible. Ask the adults you live with if more insulation could be added to your home or if an insulated jacket could be put on the water heater.

Energy from Atoms

Most electricity in the United States is generated in power plants that use fossil fuels. However, alternate sources of energy exist. **Nuclear energy** is an alternate energy source produced from atomic reactions. When the nucleus of a heavy element is split, lighter elements form and energy is released. This energy can be used to light a home or power the submarines shown in **Figure 8**.

The splitting of heavy elements to produce energy is called nuclear fission. Nuclear fission is carried out in nuclear power plants using a type of uranium as fuel.

Practicing Energy Conservation

Procedure
1. Have an adult help you find the **electric meter** for your home and record the reading in your **Science Journal**.
2. Do this for several days, taking your meter readings at about the same time each day.
3. List things you and your family can do to reduce your electricity use.
4. Encourage your family to try some of the listed ideas for several days.

Analysis
1. Keep taking meter readings and infer whether the changes make any difference.
2. Have you and your family helped conserve energy?

Figure 8 Atoms can be a source of energy.

These submarines are powered by nuclear fission.

During nuclear fission, energy is given off when a heavy atom, like uranium, splits into lighter atoms.

Heavy atom

Lighter atoms

+ Energy

Figure 9 Heat released in nuclear reactors produces steam, which in turn is used to produce electricity. This is an example of transforming nuclear energy into electrical energy.
Infer Why do you think nuclear power plants are located near rivers and lakes?

Electricity from Nuclear Energy A nuclear power plant, shown in **Figure 9,** has a large chamber called a nuclear reactor. Within the nuclear reactor, uranium fuel rods sit in a pool of cooling water. Neutrons are fired into the fuel rods. When the uranium-235 atoms are hit, they break apart and fire out neutrons that hit other atoms, beginning a chain reaction. As each atom splits, it not only fires neutrons but also releases heat that is used to boil water to make steam. The steam drives a turbine, which turns a generator that produces electricity.

Reading Check *How is nuclear energy used to produce electricity?*

Nuclear energy from fission is considered to be a nonrenewable energy resource because it uses uranium-235 as fuel. A limited amount of uranium-235 is available for use. Another problem with nuclear energy is the waste material that it produces. Nuclear waste from power plants consists of highly radioactive elements formed by the fission process. Some of this waste will remain radioactive for thousands of years. The Environmental Protection Agency (EPA) has determined that nuclear waste must be stored safely and contained for at least 10,000 years before reentering the environment.

Fusion Environmental problems related to nuclear power could be eliminated if usable energy could be obtained from fusion. The Sun is a natural fusion power plant that provides energy for Earth and the solar system. Someday fusion also might provide energy for your home.

During fusion, materials of low mass are fused together to form a substance of higher mass. No fuel problem exists if the low-mass material is a commonly occurring substance. Also, if the end product is not radioactive, storing nuclear waste is not a problem. In fact, fusion of hydrogen into helium would satisfy both of these conditions. However, technologies do not currently exist to enable humans to fuse hydrogen into helium at reasonably low temperatures in a controlled manner. But research is being conducted, as shown in **Figure 10.** If this is accomplished, nuclear energy could be considered an inexhaustible fuel resource. You will learn the importance of inexhaustible and renewable energy resources in the next section.

Figure 10 Lasers are used in research facilities to help people understand and control fusion.

section 1 review

Summary

Fossil Fuels
- Coal, natural gas, and oil are all nonrenewable energy sources.
- Synthetic fuels are human-made fuels that can be derived from coal.
- The four stages of coal formation are peat, lignite, bituminous coal, and anthracite coal.
- Oil and gas are made from the decay of ancient marine organisms.
- Strip mining and underground mining are two common methods that are used to extract coal reserves.

Energy from Atoms
- Energy is released during a fission reaction when a heavy atom is split into lighter atoms.
- Fusion occurs when two atoms come together to form a single atom.

Self Check

1. **Explain** why coal, oil, and natural gas are fossil fuels.
2. **Explain** why fossil fuels are considered to be nonrenewable energy resources.
3. **Describe** two disadvantages of nuclear energy.
4. **Think Critically** Why are you likely to find natural gas and oil deposits in the same location, but less likely to find coal and petroleum deposits at the same location?

Applying Math

5. **Design a Graph** Current energy consumption by source in the U.S. is as follows: oil, 39%; natural gas, 24%; coal, 23%; nuclear energy, 8%; renewable resources, 6%. Design a bar graph to show the energy consumption by source in the U.S. Display the sources from greatest to least.

section 2
Renewable Energy Resources

as you read

What You'll Learn
- **Compare and contrast** inexhaustible and renewable energy resources.
- **Explain** why inexhaustible and renewable resources are used less than nonrenewable resources.

Why It's Important
As fossil fuel reserves continue to diminish, alternate energy resources will be needed.

Review Vocabulary
energy: the ability to cause change

New Vocabulary
- solar energy
- wind farm
- hydroelectric energy
- geothermal energy
- biomass energy

Inexhaustible Energy Resources

How soon the world runs out of fossil fuels depends on how they are used and conserved. Fortunately, there are inexhaustible energy resources. These sources of renewable energy are constant and will not run out in the future. Inexhaustible energy resources include the Sun, wind, water, and geothermal energy.

Energy from the Sun When you sit in the Sun, walk into the wind, or sail against an ocean current, you are experiencing the power of solar energy. **Solar energy** is energy from the Sun. You already know that the Sun's energy heats Earth, and it causes circulation in Earth's atmosphere and oceans. Global winds and ocean currents are examples of nature's use of solar energy. Thus, solar energy is used indirectly when the wind and some types of moving water are used to do work.

People can use solar energy in a passive way or in an active way. South-facing windows on buildings act as passive solar collectors, warming exposed rooms. Solar cells actively collect energy from the Sun and transform it into electricity. Solar cells were invented to generate electricity for satellites. Now they also are used to power calculators, streetlights, and experimental cars. Some people have installed solar energy cells on their roofs, as shown in **Figure 11.**

Figure 11 Solar panels, such as on this home in Laguna Niguel, California, can be used to collect inexhaustible solar energy to power appliances and heat water.

Figure 12 Wind farms are used to produce electricity.
Evaluate Some people might argue that windmills produce visual pollution. Why do you think this is?

Disadvantages of Solar Energy Solar energy is clean and inexhaustible, but it does have some disadvantages. Solar cells work less efficiently on cloudy days and cannot work at all at night. Some systems use batteries to store solar energy for use at night or on cloudy days, but it is difficult to store large amounts of energy in batteries. Worn out batteries also must be discarded. This can pollute the environment if not done properly.

Energy from Wind What is better to do on a warm, windy day than fly a kite? A strong wind can lift a kite high in the sky and whip it around. The pull of the wind is so great that you wonder if it will whip the kite right out of your hands. Wind is a source of energy. It was and still is used to power sailing ships. Windmills have used wind energy to grind corn and pump water. Today, windmills can be used to generate electricity. When a large number of windmills are placed in one area for the purpose of generating electricity, the area is called a **wind farm,** as shown in **Figure 12.**

Wind energy has advantages and disadvantages. Wind is nonpolluting and free. It does little harm to the environment and produces no waste. However, only a few regions of the world have winds strong enough to generate electricity. Also, wind isn't steady. Sometimes it blows too hard and at other times it is too weak or stops entirely. For an area to use wind energy consistently, the area must have a persistent wind that blows at an appropriate speed.

Physicists The optimal speed of wind needed to rotate blades on a windmill is something a physicist would study. They can calculate the energy produced based on the speed at which the blades turn. Some areas in the country are better suited for wind farms than others. Find out which areas utilize wind farms and report in your Science Journal how much electric-ity is produced and what it is used for. What kinds of organizations would a physicist work for in these locations?

 Why are some regions better suited for wind farms than others?

Energy from Water For a long time, waterwheels steadily spun next to streams and rivers. The energy in the flowing water powered the wheels that ground grain or cut lumber. More than a pretty picture, using a waterwheel in this way is an example of microhydropower. Microhydropower has been used throughout the world to do work.

Running water also can be used to generate electricity. Electricity produced by waterpower is called **hydroelectric energy.** To generate electricity from water running in a river, a large concrete dam is built to retain water, as illustrated in **Figure 13.** A lake forms behind the dam. As water is released, its force turns turbines at the base of the dam. The turbines then turn generators that make electricity.

At first it might appear that hydroelectric energy doesn't create any environmental problems and that the water is used with little additional cost. However, when dams are built, upstream lakes fill with sediment and downstream erosion increases. Land above the dam is flooded, and wildlife habitats are damaged.

Energy from Earth Erupting volcanoes and geysers like Old Faithful are examples of geothermal energy in action. The energy that causes volcanoes to erupt or water to shoot up as a geyser also can be used to generate electricity. Energy obtained by using hot magma or hot, dry rocks inside Earth is called **geothermal energy.**

Bodies of magma can heat large reservoirs of groundwater. Geothermal power plants use steam from the reservoirs to produce electricity, as shown in **Figure 14.** In a developing method, water becomes steam when it is pumped through broken, hot, dry rocks. The steam then is used to turn turbines that run generators to make electricity. The advantage of using hot, dry rocks is that they are found just about everywhere. Geothermal energy presently is being used in Hawaii and in parts of the western United States.

Figure 13 Hydroelectric power is important in many regions of the United States. Hoover Dam was built on the Colorado River to supply electricity for a large area.

The power of running water is converted to usable energy in a hydroelectric power plant.

Figure 14 Geothermal energy is used to supply electricity to industries and homes.

What by-product is produced in this geothermal plant in California? Is it considered a pollutant?

Renewable Energy Resources

Energy resources that can be replaced in nature or by humans within a relatively short period of time are referred to as renewable energy resources. This short period of time is defined generally as being within a human life span. For example, trees can be considered a renewable energy resource. As one tree is cut down, another can be planted in its place. The new tree might be left untouched or harvested.

You have learned that most energy used in the United States—about 90 percent—comes from fossil fuels, which are nonrenewable energy resources. Next, you'll look at some renewable energy resources and how they might fit into the world's total energy needs now and in the future.

Biomass Energy

A major renewable energy resource is biomass materials. **Biomass energy** is energy derived from burning organic material such as wood, alcohol, and garbage. The term *biomass* is derived from the words *biological* and *mass*.

Topic: Biomass Energy
Visit bookf.msscience.com for Web links to information about biomass energy.

Activity List three new technologies that turn biomass into useable energy. Give two examples of each type of biomass and its energy technology.

Figure 15 These campers are using wood, a renewable energy resource, to produce heat and light.
Discuss Why do you think wood is the most commonly used biomass fuel?

Energy from Wood If you've ever sat around a campfire, like the campers shown in **Figure 15,** or close to a wood-burning fireplace to keep warm, you have used energy from wood. The burning wood is releasing stored solar energy as heat energy. Humans have long used wood as an energy resource. Much of the world still cooks with wood. In fact, firewood is used more widely today than any other type of biomass fuel.

Using wood as a biomass fuel has its problems. Gases and small particles are released when wood is burned. These materials can pollute the air. When trees are cut down for firewood, natural habitats are destroyed. However, if proper conservation methods are employed or if tree farms are maintained specifically for use as fuel, energy from wood can be a part of future energy resources.

Energy from Alcohol Biomass fuel can be burned directly, such as when wood or peat is used for heating and cooking. However, it also can be transformed into other materials that might provide cleaner, more efficient fuels.

For example, during distillation, biomass fuel, such as corn, is changed to an alcohol such as ethanol. Ethanol then can be mixed with another fuel. When the other fuel is gasoline, the mixture is called gasohol. Gasohol can be used in the same way as gasoline, as shown in **Figure 16,** but it cuts down on the amount of fossil fuel needed to produce gasoline. Fluid biomass fuels are more efficient and have more uses than solid biomass fuels do.

The problem with this process is that presently, growing the corn and distilling the ethanol often uses more energy from burning fossil fuels than the amount of energy that is derived from burning ethanol. At present, biomass fuel is best used locally.

Reading Check What are the drawbacks of biomass fuels?

Figure 16 Gasohol sometimes is used to reduce dependence on fossil fuels.

Energy from Garbage Every day humans throw away a tremendous amount of burnable garbage. As much as two thirds of what is thrown away could be burned. If more garbage were used for fuel, as shown in **Figure 17,** human dependence on fossil fuels would decrease. Burning garbage is a cheap source of energy and also helps reduce the amount of material that must be dumped into landfills.

Compared to other nations, the United States lags in the use of municipal waste as a renewable energy resource. For example, in some countries in Western Europe, as much as half of the waste generated is used for biomass fuel. When the garbage is burned, heat is produced, which turns water to steam. The steam turns turbines that run generators to produce electricity.

Unfortunately, some problems can be associated with using energy from garbage. Burning municipal waste can produce toxic ash residue and air pollution. Substances such as heavy metals could find their way into the smoke from garbage and thus into the atmosphere.

Figure 17 Garbage can be burned to produce electricity at trash-burning power plants such as this one in Virginia.

section 2 review

Summary

Inexhaustible Energy Resources
- Solar cells are used to collect the Sun's energy.
- Wind energy produces no waste or pollution, however only a few areas are conducive for creating significant energy supplies.
- Dams are used to help provide running water, which is used to produce electricity.
- Energy obtained by using heat from inside Earth is called geothermal energy.

Renewable Energy Resources
- Biomass energy is produced when organic material such as wood, alcohol, or garbage is burned.
- Trash-burning power plants convert waste into electricty by burning garbage.

Self Check

1. **List** three advantages and disadvantages of using solar energy, wind energy, and hydroelectric energy.
2. **Explain** the difference between inexhaustible and renewable energy resources. Give two examples of each.
3. **Describe** how geothermal energy is used to create electricity.
4. **Infer** why nonrenewable resources are used more than inexhaustible and renewable resources.
5. **Think Critically** How could nuclear energy, which normally is classified as a nonrenewable energy resource, be reclassified as an inexhaustible energy resource?

Applying Skills

6. **Use a Spreadsheet** Make a table of energy resources. Include an example of how each resource is used. Then describe how you could reduce the use of energy resources at home.

Soaking Up Solar Energy

Winter clothing tends to be darker in color than summer clothing. The color of the material used in the clothing affects its ability to absorb energy. In this lab, you will use different colors of soil to study this effect.

Time and Temperature

Time (min)	Temperature Dish A (°C)	Temperature Dish B (°C)	Temperature Dish C (°C)
0.0			
0.5	Do not write in this book.		
1.0			
1.5			

Real-World Question
How does color affect the absorption of energy?

Goals
- **Determine** whether color has an effect on the absorption of solar energy.
- **Relate** the concept of whether color affects absorption to other applications.

Materials
dry, black soil
dry, brown soil
dry, sandy, white soil
thermometers (3)
ring stand
graph paper
colored pencils (3)
metric ruler
clear-glass or plastic dishes (3)
200-watt gooseneck lamp
*200-watt lamp with reflector and clamp
watch or clock with second hand
*stopwatch
*Alternate materials

Safety Precautions

WARNING: *Handle glass with care so as not to break it. Wear thermal mitts when handling the light source.*

Procedure
1. Fill each dish with a different color of soil to a depth of 2.5 cm.
2. Arrange the dishes close together on your desk and place a thermometer in each dish. Be sure to cover the thermometer bulb in each dish completely with the soil.
3. Position the lamp over all three dishes.
4. **Design** a data table for your observations similar to the sample table above. You will need to read the temperature of each dish every 30 s for 20 min after the light is turned on.
5. Turn on the light and begin your experiment.
6. Use the data to construct a graph. Time should be plotted on the horizontal axis and temperature on the vertical axis. Use a different colored pencil to plot the data for each type of soil, or use a computer to design a graph that illustrates your data.

Conclude and Apply
1. **Observe** which soil had the greatest temperature change. The least?
2. **Explain** why the curves on the graph flatten.
3. **Infer** Why do flat-plate solar collectors have black plates behind the water pipes?
4. **Explain** how the color of a material affects its ability to absorb energy.
5. **Infer** Why is most winter clothing darker in color than summer clothing?

section 3
Mineral Resources

Metallic Mineral Resources

If your room at home is anything like the one shown in **Figure 18,** you will find many metal items. Metals are obtained from Earth materials called metallic mineral resources. A **mineral resource** is a deposit of useful minerals. See how many metals you can find. Is there anything in your room that contains iron? What about the metal in the frame of your bed? Is it made of iron? If so, the iron might have come from the mineral hematite. What about the framing around the windows in your room? Is it aluminum? Aluminum, like that in a soft-drink can, comes from a mixture of minerals known as bauxite. Many minerals contain these and other useful elements. Which minerals are mined as sources for the materials you use every day?

Ores Deposits in which a mineral or minerals exist in large enough amounts to be mined at a profit are called **ores.** Generally, the term ore is used for metallic deposits, but this is not always the case. The hematite that was mentioned earlier as an iron ore and the bauxite that was mentioned earlier as an aluminum ore are metallic ores.

Reading Check *What is an ore?*

as you read

***What* You'll Learn**
- **Explain** the conditions needed for a mineral to be classified as an ore.
- **Describe** how market conditions can cause a mineral to lose its value as an ore.
- **Compare and contrast** metallic and nonmetallic mineral resources.

***Why* It's Important**
Many products you use are made from mineral resources.

Review Vocabulary
metal: a solid material that is generally hard, shiny, pliable and a good electrical conductor

New Vocabulary
• mineral resource
• ore
• recycling

Copper in wires found in electrical equipment comes from the mineral chalcopyrite.

Many bed frames contain iron, which is extracted from minerals such as hematite.

Aluminum comes from a mixture of minerals called bauxite.

Stainless steel contains chromium, which comes from the mineral chromite.

Figure 18 Many items in your home are made from metals obtained from metallic mineral resources.

Figure 19 Iron ores are smelted to produce nearly pure iron. **List** *three examples of what this iron could be used for.*

Economic Effects When is a mineral deposit considered an ore? The mineral in question must be in demand. Enough of it must be present in the deposit to make it worth removing. Some mining operations are profitable only if a large amount of the mineral is needed. It also must be fairly easy to separate the mineral from the material in which it is found. If any one of these conditions isn't met, the deposit might not be considered an ore.

Supply and demand is an important part of life. You might have noticed that when the supply of fresh fruit is down, the price you pay for it at the store goes up. Economic factors largely determine what an ore is.

Refining Ore The process of extracting a useful substance from an ore involves two operations—concentrating and refining. After a metallic ore is mined from Earth's crust, it is crushed and the waste rock is removed. The waste rock that must be removed before a mineral can be used is called gangue (GANG).

INTEGRATE Chemistry Refining produces a pure or nearly pure substance from ore. For example, iron can be concentrated from the ore hematite, which is composed of iron oxide. The concentrated ore then is refined to be as close to pure iron as possible. One method of refining is smelting, illustrated in **Figure 19.** Smelting is a chemical process that removes unwanted elements from the metal that is being processed. During one smelting process, a concentrated ore of iron is heated with a specific chemical. The chemical combines with oxygen in the iron oxide, resulting in pure iron. Note that one resource, fossil fuel, is burned to produce the heat that is needed to obtain the finished product of another resource, in this case iron.

Nonmetallic Mineral Resources

Any mineral resources not used as fuels or as sources of metals are nonmetallic mineral resources. These resources are mined for the nonmetallic elements contained in them and for the specific physical and chemical properties they have. Generally, nonmetallic mineral resources can be divided into two different groups—industrial minerals and building materials. Some materials, such as limestone, belong to both groups of nonmetallic mineral resources, and others are specific to one group or the other.

Industrial Minerals Many useful chemicals are obtained from industrial minerals. Sandstone is a source of silica (SiO_2), which is a compound that is used to make glass. Some industrial minerals are processed to make fertilizers for farms and gardens. For example, sylvite, a mineral that forms when seawater evaporates, is used to make potassium fertilizer.

Many people enjoy a little sprinkle of salt on french fries and pretzels. Table salt is a product derived from halite, a nonmetallic mineral resource. Halite also is used to help melt ice on roads and sidewalks during winter and to help soften water.

Other industrial minerals are useful because of their characteristic physical properties. For example, abrasives are made from deposits of corundum and garnet. Both of these minerals are hard and able to scratch most other materials they come into contact with. Small particles of garnet can be glued onto a sheet of heavy paper to make abrasive sandpaper. **Figure 20** illustrates just a few ways in which nonmetallic mineral resources help make your life more convenient.

Mini LAB

Observing the Effects of Insulation

Procedure
1. Pour **warm water** into a **thermos bottle**. Cap it and set it aside.
2. Pour **cold water** with **ice** into a **glass** surrounded by a **thermal cup holder**.
3. Pour warm water—the same temperature as in step 1—into an **uncovered cup**. Pour cold water with ice into a glass container that is not surrounded by a thermal cup holder.
4. After 2 h, measure the temperature of each of the liquids.

Analysis
1. Infer how the insulation affected the temperature of each liquid.
2. Relate the usefulness of insulation in a thermos bottle to the usefulness of fiberglass insulation in a home.

Figure 20 You benefit from the use of industrial minerals every day.

Road salt melts ice on streets.

Many important chemicals are made from industrial minerals.

An industrial mineral called trona is important for making glass.

Building Materials One of the most important nonmetallic mineral resources is aggregate. Aggregate is composed of crushed stone or a mixture of gravel and sand and has many uses in the building industry. For example, aggregates can be mixed with cement and water to form concrete. Quality concrete is vital to the building industry. Limestone also has industrial uses. It is used as paving stone and as part of concrete mixtures. Have you ever seen the crushed rock in a walking path or driveway? The individual pieces might be crushed limestone. Gypsum, a mineral that forms when seawater evaporates, is soft and lightweight and is used in the production of plaster and wallboard. If you handle a piece of broken plaster or wallboard, note its appearance, which is similar to the mineral gypsum.

Rock also is used as building stone. You might know of buildings in your region that are made from granite, limestone, or sandstone. These rocks and others are quarried and cut into blocks and sheets. The pieces then can be used to construct buildings. Some rock also is used to sculpt statues and other pieces of art.

Reading Check *What are some important nonmetallic mineral resources?*

Applying Science

Why should you recycle?

Recycling in the United States has become a way of life. In 2000, 88 percent of Americans participated in recycling. Recycling is important because it saves precious raw materials and energy. Recycling aluminum saves 95 percent of the energy required to obtain it from its ore. Recycling steel saves up to 74 percent in energy costs, and recycling glass saves up to 22 percent.

Recycling Rates in the United States

Material	1995 (%)	1997 (%)	2001 (%)
Glass	24.5	24.3	27.2
Steel	36.5	38.4	43.5 (est.)
Aluminum	34.6	31.2	33.0
Plastics	5.3	5.2	7.0

Identifying the Problem

The following table includes materials that currently are being recycled and rates of recycling for the years 1995, 1997, and 2001. Examine the table to determine materials for which recycling increased or decreased between 1995 and 2001.

Solving the Problem

1. Has the recycling of materials increased or decreased over time? Which materials are recycled most? Which materials are recycled least? Discuss why some materials might be recycled more than others.
2. How can recycling benefit society? Explain your answer.

Recycling Mineral Resources

Mineral resources are nonrenewable. You've learned that nonrenewable resources are those that Earth processes cannot replace within an average human's lifetime. Most mineral resources take millions of years to form. Have you ever thrown away an empty soft-drink can? Many people do. These cans become solid waste. Wouldn't it be better if these cans and other items made from mineral resources were recycled into new items?

Recycling is using old materials to make new ones. Recycling has many advantages. It reduces the demand for new mineral resources. The recycling process often uses less energy than it takes to obtain new material. Because supplies of some minerals might become limited in the future, recycling could be required to meet needs for certain materials, as shown in **Figure 21**.

Recycling also can be a profitable experience. Some companies purchase scrap metal and empty soft-drink cans for the aluminum and tin content. The seller receives a small amount of money for turning in the material. Schools and other groups earn money by recycling soft-drink cans.

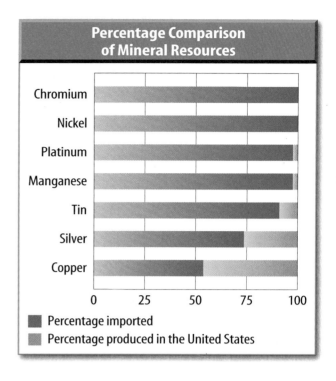

Figure 21 The United States produces only a small percentage of the metallic resources it consumes.

section 3 review

Summary

Metallic Mineral Resources
- Minerals found in rocks that can be mined for an economic profit are called ores.

Nonmetallic Mineral Resources
- Nonmetallic mineral resources can be classified into groups: industrial minerals and building materials.
- Sedimentary rocks such as limestone and sandstone can be used as building materials to make things like buildings and statues.

Recycling Mineral Resources
- Recycling materials helps to preserve Earth's resources by reusing old or used materials without extracting new resources from Earth.
- The recycling process may use fewer resources than it takes to obtain new material.

Self Check

1. **Explain** how metals obtained from metallic mineral resources are used in your home and school. Which of these products could be recycled easily?
2. **List** two industrial uses for nonmetallic mineral resources.
3. **Explain** how supply and demand of a material can cause a mineral to become an ore.
4. **Think Critically** Gangue is waste rock remaining after a mineral ore is removed. Why is gangue sometimes reprocessed?

Applying Skills

5. **Classify** the following mineral resources as metallic or nonmetallic: *hematite, limestone, bauxite, sandstone, garnet,* and *chalcopyrite*. Explain why you classified each one as you did.

LAB Model and Invent

Home Sweet Home

Goals
- **Research** various inexhaustible and other energy resources available to use in the home.
- **Design** blueprints for an energy-efficient home and/or design and build a model of an energy-efficient home.

Possible Materials
paper
ruler
pencils
cardboard
glue
aluminum foil

Real-World Question

As fossil fuel supplies continue to be depleted, an increasing U. S. population has recognized the need for alternative energy sources. United States residents might be forced to consider using inexhaustible and other renewable energy resources to meet some of their energy needs. The need for energy-efficient housing is more relevant now than ever before. A designer of energy-efficient homes considers proper design and structure, a well chosen building site with wise material selection, and selection of efficient energy generation systems to power the home. Energy-efficient housing uses less energy and produces fewer pollutants. What does the floor plan, building plan, or a model of an energy efficient home look like? How and where should your house be designed and built to efficiently use the alternative energy resources you've chosen?

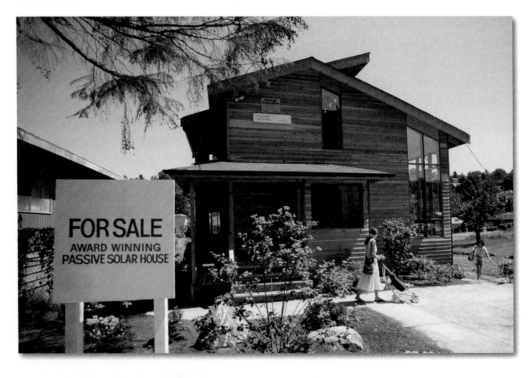

Using Scientific Methods

▶ Make a Model

Plan

1. **Research** current information about energy-efficient homes.
2. **Research** inexhaustible energy resources such as wind, hydroelectric power, or solar power, as well as energy conservation. Decide which energy resources are most efficient for your home design.
3. Decide where your house should be built to use energy efficiently.
4. Decide how your house will be laid out and draw mock blueprints for your home. Highlight energy issues such as where solar panels can be placed.
5. Build a model of your energy-efficient home.

Do

1. Ask your peers for input on your home. As you research, become an expert in one area of alternative energy generation and share your information with your classmates.
2. **Compare** your home's design to energy-efficient homes you learn about through your research.

▶ Test Your Model

1. Think about how most of the energy in a home is used. Remember as you plan your home that energy-efficient homes not only generate energy—they also use it more efficiently.
2. Carefully consider where your home should be built. For instance, if you plan to use wind power, will your house be built in an area that receives adequate wind?
3. Be sure to plan for backup energy generation. For instance, if you plan to use mostly solar energy, what will you do if it's a cloudy day?

▶ Analyze Your Data

Devise a budget for building your home. Could your energy-efficient home be built at a reasonable price? Could anyone afford to build it?

▶ Conclude and Apply

Create a list of pro and con statements about the use of energy-efficient homes. Why aren't inexhaustible and other renewable energy sources widely used in homes today?

Communicating Your Data

Present your model to the class. Explain which energy resources you chose to use in your home and why. Have an open house. Take prospective home owners/classmates on a tour of your home and sell it.

Oops! Accidents in SCIENCE

SOMETIMES GREAT DISCOVERIES HAPPEN BY ACCIDENT!

BLACK GOLD!

What if you went out to your backyard, started digging a hole, and all of the sudden oil spurted out of the ground? Dollar signs might flash before your eyes.

It wasn't quite that exciting for Charles Tripp. Tripp, a Canadian, is credited with being the first person to strike oil. And he wasn't even looking for what has become known as "black gold."

In 1851, Tripp built a factory in Ontario, Canada, not far from Lake Erie. He used a natural, black, thick, sticky substance that could be found nearby to make asphalt for paving roads and to construct buildings.

In 1855, Tripp dug a well looking for fresh-water for his factory. After digging just 2 m or so, he unexpectedly came upon liquid. It wasn't clear, clean, and delicious; it was smelly thick, and black. You guessed it—oil! Tripp didn't understand the importance of his find. Two years after his accidental discovery, Tripp sold his company to James Williams. In 1858,

Some people used TNT to search for oil. This photo was taken in 1943.

Williams continued to search for water for the factory, but, as luck would have it, diggers kept finding oil.

Some people argue that the first oil well in North America was in Titusville, Pennsylvania, when Edwin Drake hit oil in 1859. However, most historians agree that Williams was first in 1858. But they also agree that it was Edwin Drake's discovery that led to the growth of the oil industry. So, Drake and Williams can share the credit!

The Titusville, Pennsylvania, oil well drilled by Edwin Drake. This photo was taken in 1864.

Today, many oil companies are drilling beneath the sea for oil.

Make a Graph Research the leading oil-producing nations and make a bar graph of the top five producers. Research how prices of crude oil affect the U.S. and world economies. Share your findings with your class.

For more information, visit bookf.msscience.com/oops

Chapter 3 Study Guide

Reviewing Main Ideas

Section 1 Nonrenewable Energy Resources

1. Fossil fuels are considered to be non-renewable energy resources.
2. The higher the concentration of carbon in coal is, the cleaner it burns.
3. Oil and natural gas form from altered and buried marine organisms and often are found near one another.
4. Nuclear energy is obtained from the fission of heavy isotopes.

Section 2 Renewable Energy Resources

1. Inexhaustible energy resources—solar energy, wind energy, hydroelectric energy, and geothermal energy—are constant and will not run out.
2. Renewable energy resources are replaced within a relatively short period of time.
3. Biomass energy is derived from organic material such as wood and corn.

Section 3 Mineral Resources

1. Metallic mineral resources provide metals.
2. Ores are mineral resources that can be mined at a profit.
3. Smelting is a chemical process that removes unwanted elements from a metal that is being processed.
4. Nonmetallic mineral resources are classified as industrial minerals or building materials.

Visualizing Main Ideas

Copy and complete the following table that lists advantages and disadvantages of energy resources.

Energy Resources		
Resource	Advantages	Disadvantages
Fossil fuels		
Nuclear energy		
Solar energy	Do not write in this book.	
Wind energy		
Geothermal energy		
Biomass fuel		

chapter 3 Review

Using Vocabulary

biomass energy p. 79
coal p. 67
fossil fuel p. 66
geothermal energy p. 78
hydroelectric energy p. 78
mineral resource p. 83
natural gas p. 69
nuclear energy p. 73
oil p. 69
ore p. 83
recycling p. 87
reserve p. 71
solar energy p. 76
wind farm p. 77

Each phrase below describes a vocabulary word from the list. Write the word that matches the phrase describing it.

1. mineral resource mined at a profit
2. fuel that is composed mainly of the remains of dead plants
3. method of conservation in which items are processed to be used again
4. inexhaustible energy resource that is used to power the *Hubble Space Telescope*
5. energy resource that is based on fission
6. liquid from remains of marine organisms

Checking Concepts

Choose the word or phrase that best answers the question.

7. Which has the highest content of carbon?
   A) peat
   B) lignite
   C) bituminous coal
   D) anthracite coal

8. Which is the first step in coal formation?
   A) peat
   B) lignite
   C) bituminous coal
   D) anthracite

9. Which of the following is an example of a fossil fuel?
   A) wind
   B) water
   C) natural gas
   D) uranium-235

10. What is the waste material that must be separated from an ore?
    A) smelter
    B) gangue
    C) mineral resource
    D) petroleum

11. What common rock structure can trap oil and natural gas under it?
    A) folded rock
    B) sandstone rock
    C) porous rock
    D) permeable rock

Use the figure below to answer question 12.

Heavy atom   Lighter atoms

12. What other particles are released in the reaction above?
    A) protons
    B) neutrons
    C) uranium atoms
    D) heavy atoms

13. What is a region where many windmills are located in order to generate electricity from wind called?
    A) wind farm
    B) hydroelectric dam
    C) oil well
    D) steam-driven turbine

14. Which of the following is a deposit of hematite that can be mined at a profit?
    A) ore
    B) anthracite
    C) gangue
    D) energy resource

15. What is an important use of petroleum?
    A) making plaster
    B) making glass
    C) as abrasives
    D) making gasoline

16. Which of the following is a nonrenewable energy resource?
    A) water
    B) wind
    C) geothermal
    D) petroleum

chapter 3 Review

Thinking Critically

17. **Describe** the major problems associated with generating electricity using nuclear power plants.

18. **Explain** why wind is considered to be an inexhaustible energy resource.

19. **Determine** which type of energy resources are considered to be biomass fuels. List three biomass fuels.

20. **Discuss** two conditions which could occur to cause gangue to be reclassified as an ore.

21. **Predict** If a well were drilled into a rock layer containing petroleum, natural gas, and water, which substance would be encountered first? Illustrate your answer with a labeled diagram.

22. **Compare and contrast** solar energy and wind energy by creating a table.

23. **Concept Map** Copy and complete the following concept map about mineral resources.

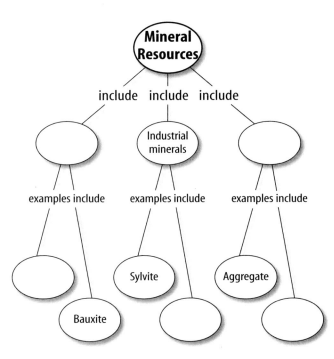

Performance Activities

24. **Make Models** Make a blueprint of a house that has been built to use passive solar energy. Which side of the house will face the sun?

25. **Letter** Write a letter to the Department of Energy asking how usable energy might be obtained from methane hydrates in the future. Also inquire about methods to extract methane hydrates.

Applying Math

Use the table below to answer questions 26–28.

Big Canyon Mine

Ore Mineral	Metal	Percent Composition	Value (dollars/kg)
Bauxite	Aluminum	5	1.00
Hematite	Iron	2	4.00
Chalcopyrite	Copper	1	6.00
Galena	Lead	7	1.00

26. **Ore Composition** If 100 kg of rock are extracted from this mine, what percentage of the rock is gangue?
    A. 15%   C. 74%
    B. 26%   D. 85%

27. **Total Composition** Graph the total composition of the extracted rock using a circle graph. Label each component clearly. Provide a title for your graph.

28. **Economic Geology** Of that 100 kg of extracted rock, determine how many kilograms of each ore mineral is extracted. List the total dollar value for each metal after the gangue has been eliminated and the ore mineral extracted.

chapter 3 Standardized Test Practice

Part 1 Multiple Choice

Record your answers on the answer sheet provided by your teacher or on a sheet of paper.

1. Which is a sedimentary rock formed from decayed plant matter?
   A. biomass
   B. coal
   C. natural gas
   D. oil

Use the graph below to answer questions 2 and 3.

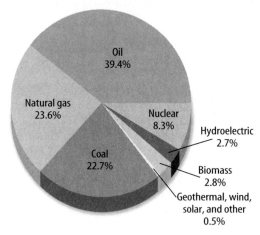

Energy Use in the United States, 2002

2. What percentage of the energy used in the United States comes from fossils fuels?
   A. 6%
   B. 24%
   C. 47%
   D. 86%

3. What percentage of our energy sources would be lost if coal runs out?
   A. 6%
   B. 8%
   C. 23%
   D. 39%

4. Which is a new potential source of methane?
   A. coal
   B. hydrates
   C. hydrocarbons
   D. petroleum

5. Which is an inexhaustible resource?
   A. coal
   B. nuclear
   C. oil
   D. solar

Use the illustration below to answer questions 6 and 7.

Heavy atom → Lighter atoms + Energy

6. Which type of energy source is shown in this diagram?
   A. coal
   B. fission
   C. fusion
   D. natural gas

7. What is produced that drives a turbine, which turns a generator?
   A. atoms
   B. neutrons
   C. steam
   D. waste

8. Which type of energy uses magma or hot dry rocks to generate electricity?
   A. geothermal
   B. hydroelectric
   C. nuclear
   D. solar

9. What is combined to make gasohol?
   A. ethanol and gasoline
   B. oil and gasoline
   C. oil and petroleum
   D. wood and gasoline

10. Which helps to reduce the demand for new mineral resources?
    A. generating
    B. mining
    C. recycling
    D. refining

Test-Taking Tip

Circle Graphs If the question asks about the sum of multiple segments of a circle graph, do your addition on scratch paper and double-check your math before selecting an answer.

Standardized Test Practice

Part 2 — Short Response/Grid In

Record your answers on the answer sheet provided by your teacher or on a sheet of paper.

11. What are two advantages of burning garbage for fuel?

12. What conditions are necessary for a wind farm?

13. Contrast methods used to remove coal with methods used to remove oil and natural gas.

14. How are bacteria involved in the formation of coal?

15. Why are methane hydrates so difficult to extract from the seafloor?

16. List two advantages of using fusion as an energy source.

17. Compare and contrast a mineral resource and an ore. How could a mineral resource become an ore? Is it possible for an ore to become just a mineral resource? Explain your answers.

Use the photo below to answer question 18.

18. What type of nonmetallic mineral resource is being used in this picture? List two other uses for this nonmetallic mineral.

Part 3 — Open Ended

Record your answers on a sheet of paper.

19. Contrast the amount of heat released and smoke produced when burning peat, lignite, bituminous coal, and anthracite coal.

20. What are some household ways to help conserve fossil fuels?

Use the illustration below to answer question 21.

21. How is energy to run the turbine being produced? Discuss environmental issues associated with this energy source.

22. Design a 4-part, time-lapse illustration to show the path of iron from the hematite mine to pure iron.

23. How do population growth and technology affect the use of nonrenewable resources?

24. Some sources describe the Sun, wind, water, and geothermal energy as inexhaustible energy resources. What might be some limitations to these resources?

25. Are mineral resources considered to be renewable or nonrenewable? Explain your answer.

chapter 4

Plate Tectonics

chapter preview

sections
1. **Continental Drift**
2. **Seafloor Spreading**
   Lab Seafloor Spreading Rates
3. **Theory of Plate Tectonics**
   Lab Predicting Tectonic Activity
   Virtual Lab Where do most earthquake epicenters and volcanoes occur?

Will this continent split?

Ol Doinyo Lengai is an active volcano in the East African Rift Valley, a place where Earth's crust is being pulled apart. If the pulling continues over millions of years, Africa will separate into two landmasses. In this chapter, you'll learn about rift valleys and other clues that the continents move over time.

Science Journal Pretend you're a journalist with an audience that assumes the continents have never moved. Write about the kinds of evidence you'll need to convince people otherwise.

Start-Up Activities

Reassemble an Image

Can you imagine a giant landmass that broke into many separate continents and Earth scientists working to reconstruct Earth's past? Do this lab to learn about clues that can be used to reassemble a supercontinent.

1. Collect interesting photographs from an old magazine.
2. You and a partner each select one photo, but don't show them to each other. Then each of you cut your photos into pieces no smaller than about 5 cm or 6 cm.
3. Trade your cut-up photo for your partner's.
4. Observe the pieces, and reassemble the photograph your partner has cut up.
5. **Think Critically** Write a paragraph describing the characteristics of the cut-up photograph that helped you put the image back together. Think of other examples in which characteristics of objects are used to match them up with other objects.

Preview this chapter's content and activities at
bookf.msscience.com

Plate Tectonics Make the following Foldable to help identify what you already know, what you want to know, and what you learned about plate tectonics.

STEP 1 Fold a vertical sheet of paper from side to side. Make the front edge about 1.25 cm shorter than the back edge.

STEP 2 Turn lengthwise and fold into thirds.

STEP 3 Unfold and cut only the layer along both folds to make three tabs.

STEP 4 Label each tab.

Identify Questions Before you read the chapter, write what you already know about plate tectonics under the left tab of your Foldable, and write questions about what you'd like to know under the center tab. After you read the chapter, list what you learned under the right tab.

section 1
Continental Drift

as you read

What You'll Learn
- **Describe** the hypothesis of continental drift.
- **Identify** evidence supporting continental drift.

Why It's Important
The hypothesis of continental drift led to plate tectonics—a theory that explains many processes in Earth.

Review Vocabulary
continent: one of the six or seven great divisions of land on the globe

New Vocabulary
- continental drift
- Pangaea

Evidence for Continental Drift

If you look at a map of Earth's surface, you can see that the edges of some continents look as though they could fit together like a puzzle. Other people also have noticed this fact. For example, Dutch mapmaker Abraham Ortelius noted the fit between the coastlines of South America and Africa more than 400 years ago.

Pangaea German meteorologist Alfred Wegener (VEG nur) thought that the fit of the continents wasn't just a coincidence. He suggested that all the continents were joined together at some time in the past. In a 1912 lecture, he proposed the hypothesis of continental drift. According to the hypothesis of **continental drift,** continents have moved slowly to their current locations. Wegener suggested that all continents once were connected as one large landmass, shown in **Figure 1,** that broke apart about 200 million years ago. He called this large landmass **Pangaea** (pan JEE uh), which means "all land."

Reading Check Who proposed continental drift?

Figure 1 This illustration represents how the continents once were joined to form Pangaea. This fitting together of continents according to shape is not the only evidence supporting the past existence of Pangaea.

98 ◆ F CHAPTER 4 Plate Tectonics

A Controversial Idea Wegener's ideas about continental drift were controversial. It wasn't until long after Wegener's death in 1930 that his basic hypothesis was accepted. The evidence Wegener presented hadn't been enough to convince many people during his lifetime. He was unable to explain exactly how the continents drifted apart. He proposed that the continents plowed through the ocean floor, driven by the spin of Earth. Physicists and geologists of the time strongly disagreed with Wegener's explanation. They pointed out that continental drift would not be necessary to explain many of Wegener's observations. Other important observations that came later eventually supported Wegener's earlier evidence.

Topic: Continental Drift
Visit bookf.msscience.com for Web links to information about the continental drift hypothesis.

Activity Research and write a brief report about the initial reactions, from the public and scientific communities, toward Wegener's continental drift hypothesis.

Fossil Clues Besides the puzzlelike fit of the continents, fossils provided support for continental drift. Fossils of the reptile *Mesosaurus* have been found in South America and Africa, as shown in **Figure 2.** This swimming reptile lived in freshwater and on land. How could fossils of *Mesosaurus* be found on land areas separated by a large ocean of salt water? It probably couldn't swim between the continents. Wegener hypothesized that this reptile lived on both continents when they were joined.

Reading Check *How do* Mesosaurus *fossils support the past existence of Pangaea?*

Figure 2 Fossil remains of plants and animals that lived in Pangaea have been found on more than one continent.
Evaluate *How do the locations of* Glossopteris, Mesosaurus, Kannemeyerid, Labyrinthodont, *and other fossils support Wegener's hypothesis of continental drift?*

Figure 3 This fossil plant, *Glossopteris,* grew in a temperate climate.

Mini LAB

Interpreting Fossil Data

Procedure
1. Build a three-layer landmass using **clay** or **modeling dough.**
2. Mold the clay into mountain ranges.
3. Place similar "**fossils**" into the clay at various locations around the landmass.
4. Form five continents from the one landmass. Also, form two smaller landmasses out of different clay with different mountain ranges and fossils.
5. Place the five continents and two smaller landmasses around the room.
6. Have someone who did not make or place the landmasses make a model that shows how they once were positioned.
7. Return the clay to its container so it can be used again.

Analysis
What clues were useful in reconstructing the original landmass?

Try at Home

A Widespread Plant Another fossil that supports the hypothesis of continental drift is *Glossopteris* (glahs AHP tur us). **Figure 3** shows this fossil plant, which has been found in Africa, Australia, India, South America, and Antarctica. The presence of *Glossopteris* in so many areas also supported Wegener's idea that all of these regions once were connected and had similar climates.

Climate Clues Wegener used continental drift to explain evidence of changing climates. For example, fossils of warm-weather plants were found on the island of Spitsbergen in the Arctic Ocean. To explain this, Wegener hypothesized that Spitsbergen drifted from tropical regions to the arctic. Wegener also used continental drift to explain evidence of glaciers found in temperate and tropical areas. Glacial deposits and rock surfaces scoured and polished by glaciers are found in South America, Africa, India, and Australia. This shows that parts of these continents were covered with glaciers in the past. How could you explain why glacial deposits are found in areas where no glaciers exist today? Wegener thought that these continents were connected and partly covered with ice near Earth's south pole long ago.

Rock Clues If the continents were connected at one time, then rocks that make up the continents should be the same in locations where they were joined. Similar rock structures are found on different continents. Parts of the Appalachian Mountains of the eastern United States are similar to those found in Greenland and western Europe. If you were to study rocks from eastern South America and western Africa, you would find other rock structures that also are similar. Rock clues like these support the idea that the continents were connected in the past.

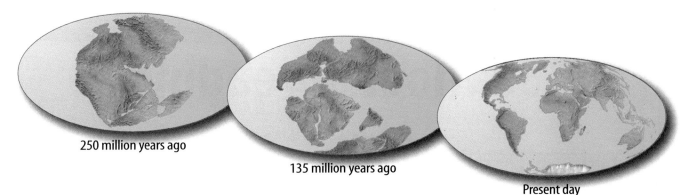

250 million years ago

135 million years ago

Present day

How could continents drift?

Although Wegener provided evidence to support his hypothesis of continental drift, he couldn't explain how, when, or why these changes, shown in **Figure 4,** took place. The idea suggested that lower-density, continental material somehow had to plow through higher-density, ocean-floor material. The force behind this plowing was thought to be the spin of Earth on its axis—a notion that was quickly rejected by physicists. Because other scientists could not provide explanations either, Wegener's idea of continental drift was initially rejected. The idea was so radically different at that time that most people closed their minds to it.

Rock, fossil, and climate clues were the main types of evidence for continental drift. After Wegener's death, more clues were found, largely because of advances in technology, and new ideas that related to continental drift were developed. You'll learn about a new idea, seafloor spreading, in the next section.

Figure 4 These computer models show the probable course the continents have taken. On the far left is their position 250 million years ago. In the middle is their position 135 million years ago. At right is their current position.

section 1 review

Summary

Evidence for Continental Drift

- Alfred Wegener proposed in his hypothesis of continental drift that all continents were once connected as one large landmass called Pangaea.
- Evidence of continental drift came from fossils, signs of climate change, and rock structures from different continents.

How could continents drift?

- During his lifetime, Wegener was unable to explain how, when, or why the continents drifted.
- After his death, advances in technology permitted new ideas to be developed to help explain his hypothesis.

Self Check

1. **Explain** how Wegener used climate clues to support his hypothesis of continental drift.
2. **Describe** how rock clues were used to support the hypothesis of continental drift.
3. **Summarize** the ways that fossils helped support the hypothesis of continental drift.
4. **Think Critically** Why would you expect to see similar rocks and rock structures on two landmasses that were connected at one time.

Applying Skills

5. **Compare and contrast** the locations of fossils of the temperate plant *Glossopteris,* as shown in **Figure 2,** with the climate that exists at each location today.

section 2
Seafloor Spreading

as you read

What You'll Learn
- **Explain** seafloor spreading.
- **Recognize** how age and magnetic clues support seafloor spreading.

Why It's Important
Seafloor spreading helps explain how continents moved apart.

Review Vocabulary
seafloor: portion of Earth's crust that lies beneath ocean waters

New Vocabulary
- seafloor spreading

Mapping the Ocean Floor

If you were to lower a rope from a boat until it reached the seafloor, you could record the depth of the ocean at that particular point. In how many different locations would you have to do this to create an accurate map of the seafloor? This is exactly how it was done until World War I, when the use of sound waves was introduced by German scientists to detect submarines. During the 1940s and 1950s, scientists began using sound waves on moving ships to map large areas of the ocean floor in detail. Sound waves echo off the ocean bottom—the longer the sound waves take to return to the ship, the deeper the water is.

Using sound waves, researchers discovered an underwater system of ridges, or mountains, and valleys like those found on the continents. In some of these underwater ridges are rather long rift valleys where volcanic eruptions and earthquakes occur from time to time. Some of these volcanoes actually are visible above the ocean surface. In the Atlantic, the Pacific, and in other oceans around the world, a system of ridges, called the mid-ocean ridges, is present. These underwater mountain ranges, shown in **Figure 5,** stretch along the center of much of Earth's ocean floor. This discovery raised the curiosity of many scientists. What formed these mid-ocean ridges?

Reading Check *How were mid-ocean ridges discovered?*

Figure 5 As the seafloor spreads apart at a mid-ocean ridge, new seafloor is created. The older seafloor moves away from the ridge in opposite directions.

The Seafloor Moves In the early 1960s, Princeton University scientist Harry Hess suggested an explanation. His now-famous theory is known as **seafloor spreading.** Hess proposed that hot, less dense material below Earth's crust rises toward the surface at the mid-ocean ridges. Then, it flows sideways, carrying the seafloor away from the ridge in both directions, as seen in **Figure 5.**

As the seafloor spreads apart, magma is forced upward and flows from the cracks. It becomes solid as it cools and forms new seafloor. As new seafloor moves away from the mid-ocean ridge, it cools, contracts, and becomes denser. This denser, colder seafloor sinks, helping to form the ridge. The theory of seafloor spreading was later supported by the following observations.

Reading Check *How does new seafloor form at mid-ocean ridges?*

Evidence for Spreading

In 1968, scientists aboard the research ship *Glomar Challenger* began gathering information about the rocks on the seafloor. *Glomar Challenger* was equipped with a drilling rig that allowed scientists to drill into the seafloor to obtain rock samples. Scientists found that the youngest rocks are located at the mid-ocean ridges. The ages of the rocks become increasingly older in samples obtained farther from the ridges, adding to the evidence for seafloor spreading.

Using submersibles along mid-ocean ridges, new seafloor features and life-forms also were discovered there, as shown in **Figure 6.** As molten material is forced upward along the ridges, it brings heat and chemicals that support exotic life-forms in deep, ocean water. Among these are giant clams, mussels, and tube worms.

Figure 6 Many new discoveries have been made on the seafloor. These giant tube worms inhabit areas near hot water vents along mid-ocean ridges.

Curie Point Find out what the Curie point is and describe in your Science Journal what happens to iron-bearing minerals when they are heated to the Curie point. Explain how this is important to studies of seafloor spreading.

Magnetic Clues Earth's magnetic field has a north and a south pole. Magnetic lines, or directions, of force leave Earth near the south pole and enter Earth near the north pole. During a magnetic reversal, the lines of magnetic force run the opposite way. Scientists have determined that Earth's magnetic field has reversed itself many times in the past. These reversals occur over intervals of thousands or even millions of years. The reversals are recorded in rocks forming along mid-ocean ridges.

■ Normal magnetic polarity
■ Reverse magnetic polarity

Figure 7 Changes in Earth's magnetic field are preserved in rock that forms on both sides of mid-ocean ridges.
Explain *why this is considered to be evidence of seafloor spreading.*

Magnetic Time Scale Iron-bearing minerals, such as magnetite, that are found in the rocks of the seafloor can record Earth's magnetic field direction when they form. Whenever Earth's magnetic field reverses, newly forming iron minerals will record the magnetic reversal.

Using a sensing device called a magnetometer (mag nuh TAH muh tur) to detect magnetic fields, scientists found that rocks on the ocean floor show many periods of magnetic reversal. The magnetic alignment in the rocks reverses back and forth over time in strips parallel to the mid-ocean ridges, as shown in **Figure 7**. A strong magnetic reading is recorded when the polarity of a rock is the same as the polarity of Earth's magnetic field today. Because of this, normal polarities in rocks show up as large peaks. This discovery provided strong support that seafloor spreading was indeed occurring. The magnetic reversals showed that new rock was being formed at the mid-ocean ridges. This helped explain how the crust could move—something that the continental drift hypothesis could not do.

section 2 review

Summary

Mapping the Ocean Floor
- Mid-ocean ridges, along the center of the ocean floor, have been found by using sound waves, the same method once used to detect submarines during World War I.
- Harry Hess suggested, in his seafloor spreading hypothesis, that the seafloor moves.

Evidence for Spreading
- Scientists aboard *Glomar Challenger* provided evidence of spreading by discovering that the youngest rocks are located at ridges and become increasingly older farther from the ridges.
- Magnetic alignment of rocks, in alternating strips that run parallel to ridges, indicates reversals in Earth's magnetic field and provides further evidence of seafloor spreading.

Self Check

1. **Summarize** What properties of iron-bearing minerals on the seafloor support the theory of seafloor spreading?
2. **Explain** how the ages of the rocks on the ocean floor support the theory of seafloor spreading.
3. **Summarize** How did Harry Hess's hypothesis explain seafloor movement?
4. **Explain** why some partly molten material rises toward Earth's surface.
5. **Think Critically** The ideas of Hess, Wegener, and others emphasize that Earth is a dynamic planet. How is seafloor spreading different from continental drift?

Applying Math

6. **Solve One-Step Equations** North America is moving about 1.25 cm per year away from a ridge in the middle of the Atlantic Ocean. Using this rate, how much farther apart will North America and the ridge be in 200 million years?

bookf.msscience.com/self_check_quiz

Seafloor Spreading Rates

How did scientists use their knowledge of seafloor spreading and magnetic field reversals to reconstruct Pangaea? Try this lab to see how you can determine where a continent may have been located in the past.

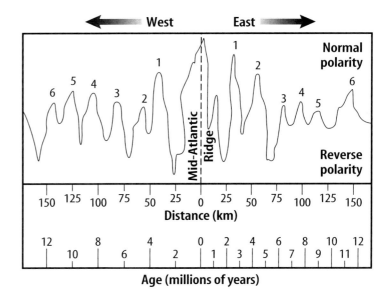

Real-World Question

Can you use clues, such as magnetic field reversals on Earth, to help reconstruct Pangaea?

Goals

- **Interpret** data about magnetic field reversals. Use these magnetic clues to reconstruct Pangaea.

Materials

metric ruler
pencil

Procedure

1. Study the magnetic field graph above. You will be working only with normal polarity readings, which are the peaks above the baseline in the top half of the graph.
2. Place the long edge of a ruler vertically on the graph. Slide the ruler so that it lines up with the center of peak 1 west of the Mid-Atlantic Ridge.
3. **Determine** and record the distance and age that line up with the center of peak 1 west. Repeat this process for peak 1 east of the ridge.
4. **Calculate** the average distance and age for this pair of peaks.
5. Repeat steps 2 through 4 for the remaining pairs of normal-polarity peaks.
6. **Calculate** the rate of movement in cm per year for the six pairs of peaks. Use the formula rate = distance/time. Convert kilometers to centimeters. For example, to calculate a rate using normal-polarity peak 5, west of the ridge:

$$\text{rate} = \frac{125 \text{ km}}{10 \text{ million years}} = \frac{12.5 \text{ km}}{\text{million years}} = \frac{1{,}250{,}000 \text{ cm}}{1{,}000{,}000 \text{ years}} = 1.25 \text{ cm/year}$$

Conclude and Apply

1. **Compare** the age of igneous rock found near the mid-ocean ridge with that of igneous rock found farther away from the ridge.
2. If the distance from a point on the coast of Africa to the Mid-Atlantic Ridge is approximately 2,400 km, calculate how long ago that point in Africa was at or near the Mid-Atlantic Ridge.
3. How could you use this method to reconstruct Pangaea?

section 3
Theory of Plate Tectonics

as you read

What You'll Learn
- **Compare and contrast** different types of plate boundaries.
- **Explain** how heat inside Earth causes plate tectonics.
- **Recognize** features caused by plate tectonics.

Why It's Important
Plate tectonics explains how many of Earth's features form.

Review Vocabulary
converge: to come together
diverge: to move apart
transform: to convert or change

New Vocabulary
- plate tectonics
- plate
- lithosphere
- asthenosphere
- convection current

Plate Tectonics

The idea of seafloor spreading showed that more than just continents were moving, as Wegener had thought. It was now clear to scientists that sections of the seafloor and continents move in relation to one another.

Plate Movements In the 1960s, scientists developed a new theory that combined continental drift and seafloor spreading. According to the theory of **plate tectonics,** Earth's crust and part of the upper mantle are broken into sections. These sections, called **plates,** move on a plasticlike layer of the mantle. The plates can be thought of as rafts that float and move on this layer.

Composition of Earth's Plates Plates are made of the crust and a part of the upper mantle, as shown in **Figure 8.** These two parts combined are the **lithosphere** (LIH thuh sfihr). This rigid layer is about 100 km thick and generally is less dense than material underneath. The plasticlike layer below the lithosphere is called the **asthenosphere** (as THE nuh sfihr). The rigid plates of the lithosphere float and move around on the asthenosphere.

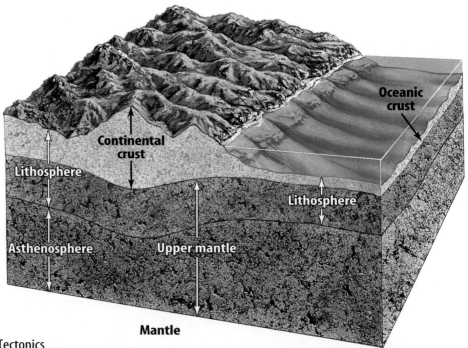

Figure 8 Plates of the lithosphere are composed of oceanic crust, continental crust, and rigid upper mantle.

Figure 9 This diagram shows the major plates of the lithosphere, their direction of movement, and the type of boundary between them. **Analyze and Conclude** Based on what is shown in this figure, what is happening where the Nazca Plate meets the Pacific Plate?

Plate Boundaries

When plates move, they can interact in several ways. They can move toward each other and converge, or collide. They also can pull apart or slide alongside one another. When the plates interact, the result of their movement is seen at the plate boundaries, as in **Figure 9**.

Reading Check *What are the general ways that plates interact?*

Movement along any plate boundary means that changes must happen at other boundaries. What is happening to the Atlantic Ocean floor between the North American and African Plates? Compare this with what is happening along the western margin of South America.

Plates Moving Apart The boundary between two plates that are moving apart is called a divergent boundary. You learned about divergent boundaries when you read about seafloor spreading. In the Atlantic Ocean, the North American Plate is moving away from the Eurasian and the African Plates, as shown in **Figure 9**. That divergent boundary is called the Mid-Atlantic Ridge. The Great Rift Valley in eastern Africa might become a divergent plate boundary. There, a valley has formed where a continental plate is being pulled apart. **Figure 10** shows a side view of what a rift valley might look like and illustrates how the hot material rises up where plates separate.

SECTION 3 Theory of Plate Tectonics **F** ◆ **107**

Science Online

Topic: Earthquakes and Volcanoes

Visit bookf.msscience.com for Web links to recent news or magazine articles about earthquakes and volcanic activity related to plate tectonics.

Activity Prepare a group demonstration about recent volcanic and earthquake events. Divide tasks among group members. Find and copy maps, diagrams, photographs, and charts to highlight your presentation. Emphasize the locations of events and the relationship to plate tectonics.

Plates Moving Together If new crust is being added at one location, why doesn't Earth's surface keep expanding? As new crust is added in one place, it disappears below the surface at another. The disappearance of crust can occur when seafloor cools, becomes denser, and sinks. This occurs where two plates move together at a convergent boundary.

When an oceanic plate converges with a less dense continental plate, the denser oceanic plate sinks under the continental plate. The area where an oceanic plate subducts, or goes down, into the mantle is called a subduction zone. Some volcanoes form above subduction zones. **Figure 10** shows how this type of convergent boundary creates a deep-sea trench where one plate bends and sinks beneath the other. High temperatures cause rock to melt around the subducting slab as it goes under the other plate. The newly formed magma is forced upward along these plate boundaries, forming volcanoes. The Andes mountain range of South America contains many volcanoes. They were formed at the convergent boundary of the Nazca and the South American Plates.

Applying Science

How well do the continents fit together?

Recall the Launch Lab you performed at the beginning of this chapter. While you were trying to fit pieces of a cut-up photograph together, what clues did you use?

Identifying the Problem

Take a copy of a map of the world and cut out each continent. Lay them on a tabletop and try to fit them together, using techniques you used in the Launch Lab. You will find that the pieces of your Earth puzzle—the continents—do not fit together well. Yet, several of the areas on some continents fit together extremely well.

Take out another world map—one that shows the continental shelves as well as the continents. Copy it and cut out the continents, this time including the continental shelves.

Solving the Problem
1. Does including the continental shelves solve the problem of fitting the continents together?
2. Why should continental shelves be included with maps of the continents?

108 ◆ F CHAPTER 4 Plate Tectonics

NATIONAL GEOGRAPHIC VISUALIZING PLATE BOUNDARIES

Figure 10

By diverging at some boundaries and converging at others, Earth's plates are continually—but gradually—reshaping the landscape around you. The Mid-Atlantic Ridge, for example, was formed when the North and South American Plates pulled apart from the Eurasian and African Plates (see globe). Some features that occur along plate boundaries—rift valleys, volcanoes, and mountain ranges—are shown on the right and below.

A RIFT VALLEY When continental plates pull apart, they can form rift valleys. The African continent is separating now along the East African Rift Valley.

SUBDUCTION Where oceanic and continental plates collide, the oceanic plate plunges beneath the less dense continental plate. As the plate descends, molten rock (yellow) forms and rises toward the surface, creating volcanoes.

SEAFLOOR SPREADING A mid-ocean ridge, like the Mid-Atlantic Ridge, forms where oceanic plates continue to separate. As rising magma (yellow) cools, it forms new oceanic crust.

CONTINENTAL COLLISION Where two continental plates collide, they push up the crust to form mountain ranges such as the Himalaya.

SECTION 3 Theory of Plate Tectonics F ◆ 109

Where Plates Collide A subduction zone also can form where two oceanic plates converge. In this case, the colder, older, denser oceanic plate bends and sinks down into the mantle. The Mariana Islands in the western Pacific are a chain of volcanic islands formed where two oceanic plates collide.

Usually, no subduction occurs when two continental plates collide, as shown in **Figure 10.** Because both of these plates are less dense than the material in the asthenosphere, the two plates collide and crumple up, forming mountain ranges. Earthquakes are common at these convergent boundaries. However, volcanoes do not form because there is no, or little, subduction. The Himalaya in Asia are forming where the Indo-Australian Plate collides with the Eurasian Plate.

Where Plates Slide Past Each Other The third type of plate boundary is called a transform boundary. Transform boundaries occur where two plates slide past one another. They move in opposite directions or in the same direction at different rates. When one plate slips past another suddenly, earthquakes occur. The Pacific Plate is sliding past the North American Plate, forming the famous San Andreas Fault in California, as seen in **Figure 11.** The San Andreas Fault is part of a transform plate boundary. It has been the site of many earthquakes.

Figure 11 The San Andreas Fault in California occurs along the transform plate boundary where the Pacific Plate is sliding past the North American Plate.

Overall, the two plates are moving in roughly the same direction.
Explain Why, then, do the red arrows show movement in opposite directions?

This photograph shows an aerial view of the San Andreas Fault.

Causes of Plate Tectonics

Many new discoveries have been made about Earth's crust since Wegener's day, but one question still remains. What causes the plates to move? Scientists now think they have a good idea. They think that plates move by the same basic process that occurs when you heat soup.

Convection Inside Earth Soup that is cooking in a pan on the stove contains currents caused by an unequal distribution of heat in the pan. Hot, less dense soup is forced upward by the surrounding, cooler, denser soup. As the hot soup reaches the surface, it cools and sinks back down into the pan. This entire cycle of heating, rising, cooling, and sinking is called a **convection current.** A version of this same process, occurring in the mantle, is thought to be the force behind plate tectonics. Scientists suggest that differences in density cause hot, plasticlike rock to be forced upward toward the surface.

Moving Mantle Material Wegener wasn't able to come up with an explanation for why plates move. Today, researchers who study the movement of heat in Earth's interior have proposed several possible explanations. All of the hypotheses use convection in one way or another. It is, therefore, the transfer of heat inside Earth that provides the energy to move plates and causes many of Earth's surface features. One hypothesis is shown in **Figure 12**. It relates plate motion directly to the movement of convection currents. According to this hypothesis, convection currents cause the movements of plates.

Mini LAB

Modeling Convection Currents

Procedure
1. Pour **water** into **a clear, colorless casserole dish** until it is 5 cm from the top.
2. Center the dish on a **hot plate** and heat it. **WARNING:** *Wear thermal mitts to protect your hands.*
3. Add a few drops of **food coloring** to the water above the center of the hot plate.
4. Looking from the side of the dish, observe what happens in the water.
5. Illustrate your observations in your **Science Journal.**

Analysis
1. Determine whether any currents form in the water.
2. Infer what causes the currents to form.

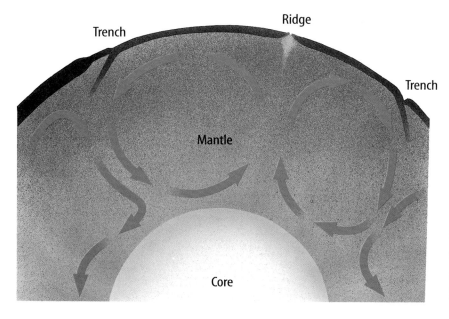

Figure 12 In one hypothesis, convection currents occur throughout the mantle. Such convection currents (see arrows) are the driving force of plate tectonics.

SECTION 3 Theory of Plate Tectonics **F** ◆ **111**

Features Caused by Plate Tectonics

Earth is a dynamic planet with a hot interior. This heat leads to convection, which powers the movement of plates. As the plates move, they interact. The interaction of plates produces forces that build mountains, create ocean basins, and cause volcanoes. When rocks in Earth's crust break and move, energy is released in the form of seismic waves. Humans feel this release as earthquakes. You can see some of the effects of plate tectonics in mountainous regions, where volcanoes erupt, or where landscapes have changed from past earthquake or volcanic activity.

Reading Check *What happens when seismic energy is released as rocks in Earth's crust break and move?*

Normal Faults and Rift Valleys Tension forces, which are forces that pull apart, can stretch Earth's crust. This causes large blocks of crust to break and tilt or slide down the broken surfaces of crust. When rocks break and move along surfaces, a fault forms. Faults interrupt rock layers by moving them out of place. Entire mountain ranges can form in the process, called fault-block mountains, as shown in **Figure 13.** Generally, the faults that form from pull-apart forces are normal faults—faults in which the rock layers above the fault move down when compared with rock layers below the fault.

Rift valleys and mid-ocean ridges can form where Earth's crust separates. Examples of rift valleys are the Great Rift Valley in Africa, and the valleys that occur in the middle of mid-ocean ridges. Examples of mid-ocean ridges include the Mid-Atlantic Ridge and the East Pacific Rise.

Figure 13 Fault-block mountains can form when Earth's crust is stretched by tectonic forces. The arrows indicate the directions of moving blocks.
Name *the type of force that occurs when Earth's crust is pulled in opposite directions.*

Mountains and Volcanoes Compression forces squeeze objects together. Where plates come together, compression forces produce several effects. As continental plates collide, the forces that are generated cause massive folding and faulting of rock layers into mountain ranges such as the Himalaya, shown in **Figure 14,** or the Appalachian Mountains. The type of faulting produced is generally reverse faulting. Along a reverse fault, the rock layers above the fault surface move up relative to the rock layers below the fault.

Reading Check *What features occur where plates converge?*

As you learned earlier, when two oceanic plates converge, the denser plate is forced beneath the other plate. Curved chains of volcanic islands called island arcs form above the sinking plate. If an oceanic plate converges with a continental plate, the denser oceanic plate slides under the continental plate. Folding and faulting at the continental plate margin can thicken the continental crust to produce mountain ranges. Volcanoes also typically are formed at this type of convergent boundary.

Volcanologist This person's job is to study volcanoes in order to predict eruptions. Early warning of volcanic eruptions gives nearby residents time to evacuate. Volcanologists also educate the public about the hazards of volcanic eruptions and tell people who live near volcanoes what they can do to be safe in the event of an eruption. Volcanologists travel all over the world to study new volcanic sites.

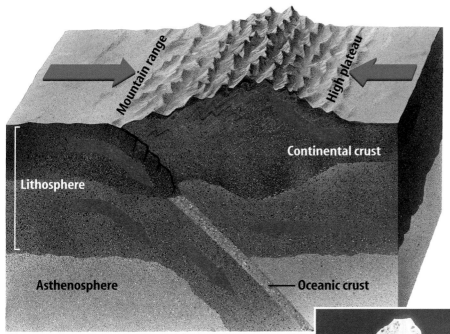

Figure 14 The Himalaya still are forming today as the Indo-Australian Plate collides with the Eurasian Plate.

Figure 15 Most of the movement along a strike-slip fault is parallel to Earth's surface. When movement occurs, human-built structures along a strike-slip fault are offset, as shown here in this road.

Strike-Slip Faults At transform boundaries, two plates slide past one another without converging or diverging. The plates stick and then slide, mostly in a horizontal direction, along large strike-slip faults. In a strike-slip fault, rocks on opposite sides of the fault move in opposite directions, or in the same direction at different rates. This type of fault movement is shown in **Figure 15.** One such example is the San Andreas Fault. When plates move suddenly, vibrations are generated inside Earth that are felt as an earthquake.

Earthquakes, volcanoes, and mountain ranges are evidence of plate motion. Plate tectonics explains how activity inside Earth can affect Earth's crust differently in different locations. You've seen how plates have moved since Pangaea separated. Is it possible to measure how far plates move each year?

Testing for Plate Tectonics

Until recently, the only tests scientists could use to check for plate movement were indirect. They could study the magnetic characteristics of rocks on the seafloor. They could study volcanoes and earthquakes. These methods supported the theory that the plates have moved and still are moving. However, they did not provide proof—only support—of the idea.

New methods had to be discovered to be able to measure the small amounts of movement of Earth's plates. One method, shown in **Figure 16,** uses lasers and a satellite. Now, scientists can measure exact movements of Earth's plates of as little as 1 cm per year.

Direction of Forces In which directions do forces act at convergent, divergent, and transform boundaries? Demonstrate these forces using wooden blocks or your hands.

Figure 16 When using the Satellite Laser Ranging System, scientists on the ground aim laser pulses at a satellite. The pulses reflect off the satellite and are used to determine a precise location on the ground.

Current Data Satellite Laser Ranging System data show that Hawaii is moving toward Japan at a rate of about 8.3 cm per year. Maryland is moving away from England at a rate of 1.7 cm per year. Using such methods, scientists have observed that the plates move at rates ranging from about 1 cm to 12 cm per year.

section 3 review

Summary

Plate Tectonics
- The theory of plate tectonics states that sections of the seafloor and continents move as plates on a plasticlike layer of the mantle.

Plate Boundaries
- The boundary between two plates moving apart is called a divergent boundary.
- Plates move together at a convergent boundary.
- Transform boundaries occur where two plates slide past one another.

Causes of Plate Tectonics
- Convection currents are thought to cause the movement of Earth's plates.

Features Caused by Plate Tectonics
- Tension forces cause normal faults, rift valleys, and mid-ocean ridges at divergent boundaries.
- At convergent boundaries, compression forces cause folding, reverse faults, and mountains.
- At transform boundaries, two plates slide past one another along strike-slip faults.

Self Check

1. **Describe** what occurs at plate boundaries that are associated with seafloor spreading.
2. **Describe** three types of plate boundaries where volcanic eruptions can occur.
3. **Explain** how convection currents are related to plate tectonics.
4. **Think Critically** Using **Figure 9** and a world map, determine what natural disasters might occur in Iceland. Also determine what disasters might occur in Tibet. Explain why some Icelandic disasters are not expected to occur in Tibet.

Applying Skills

5. **Predict** Plate tectonic activity causes many events that can be dangerous to humans. One of these events is a seismic sea wave, or tsunami. Learn how scientists predict the arrival time of a tsunami in a coastal area.
6. **Use a Word Processor** Write three separate descriptions of the three basic types of plate boundaries—divergent boundaries, convergent boundaries, and transform boundaries. Then draw a sketch of an example of each boundary next to your description.

LAB Use the Internet

Predicting Tectonic Activity

Goals
- **Research** the locations of earthquakes and volcanic eruptions around the world.
- **Plot** earthquake epicenters and the locations of volcanic eruptions.
- **Predict** locations that are tectonically active based on a plot of the locations of earthquake epicenters and active volcanoes.

Data Source

Visit bookf.msscience.com/internet_lab for more information about earthquake and volcano sites, and data from other students.

Real-World Question

The movement of plates on Earth causes forces that build up energy in rocks. The release of this energy can produce vibrations in Earth that you know as earthquakes. Earthquakes occur every day. Many of them are too small to be felt by humans, but each event tells scientists something more about the planet. Active volcanoes can do the same and often form at plate boundaries.

Can you predict tectonically active areas by plotting locations of earthquake epicenters and volcanic eruptions?

Think about where earthquakes and volcanoes have occurred in the past. Make a hypothesis about whether the locations of earthquake epicenters and active volcanoes can be used to predict tectonically active areas.

Using Scientific Methods

▸ Make a Plan

1. Make a data table in your Science Journal like the one shown.

2. Collect data for earthquake epicenters and volcanic eruptions for at least the past two weeks. Your data should include the longitude and latitude for each location. For help, refer to the data sources given on the opposite page.

Locations of Epicenters and Eruptions		
Earthquake Epicenter/ Volcanic Eruption	Longitude	Latitude
Do not write in this book.		

▸ Follow Your Plan

1. Make sure your teacher approves your plan before you start.
2. **Plot** the locations of earthquake epicenters and volcanic eruptions on a map of the world. Use an overlay of tissue paper or plastic.
3. After you have collected the necessary data, predict where the tectonically active areas on Earth are.
4. **Compare and contrast** the areas that you predicted to be tectonically active with the plate boundary map shown in **Figure 9**.

▸ Analyze Your Data

1. What areas on Earth do you predict to be the locations of tectonic activity?
2. How close did your prediction come to the actual location of tectonically active areas?

▸ Conclude and Apply

1. How could you make your predictions closer to the locations of actual tectonic activity?
2. Would data from a longer period of time help? Explain.
3. What types of plate boundaries were close to your locations of earthquake epicenters? Volcanic eruptions?
4. **Explain** which types of plate boundaries produce volcanic eruptions. Be specific.

Communicating Your Data

Find this lab using the link below. Post your data in the table provided. **Compare** your data to those of other students. Combine your data with those of other students and **plot** these combined data on a map to recognize the relationship between plate boundaries, volcanic eruptions, and earthquake epicenters.

Science online

bookf.msscience.com/internet_lab

LAB F ◆ 117

Science and Language Arts

Listening In
by Gordon Judge

I'm just a bit of seafloor on this mighty solid sphere.
With no mind to be broadened, I'm quite content
    down here.
The mantle churns below me, and the sea's in turmoil, too;
But nothing much disturbs me, I'm rock solid through
    and through.

I do pick up occasional low-frequency vibrations –
(I think, although I can't be sure, they're sperm whales'
    conversations).
I know I shouldn't listen in, but what else can I do?
It seems they are all studying for degrees from the OU.

They've mentioned me in passing, as their minds begin
    improving:

I think I've heard them say
    "The theory says the sea-
    floor's moving…".
They call it "Plate Tectonics", this
    new theory in their noddle.
If they would only ask me, I
    could tell them it's all
    twaddle….

But, how can I be moving, when I know full well myself
That I'm quite firmly anchored to a continental shelf?
"Well, the continent is moving, too; you're *pushing* it,
    you see,"
I hear those OU whales intone, hydro-acoustically….

Well, thank you very much, OU. You've upset my
    composure.
Next time you send your student whales to look at
    my exposure
I'll tell them it's a load of tosh: it's *they* who move,
    not me,
Those arty-smarty blobs of blubber, clogging up the sea!

Understanding Literature

Point of View Point of view refers to the perspective from which an author writes. This poem begins, "I'm just a bit of seafloor…." Right away, you know that the poem, or story, is being told from the point of view of the speaker, or the "first person." What effect does the first-person narration have on the story?

Respond to the Reading

1. Who is narrating the poem?
2. Why might the narrator think he or she hasn't moved?
3. **Linking Science and Writing** Using the first-person point of view, write an account from the point of view of a living or nonliving thing.

 Volcanoes can occur where two plates move toward each other. When an oceanic plate and a continental plate collide, a volcano will form. Subduction zones occur when one plate sinks under another plate. Rocks melt in the zones where these plates converge, causing magma to move upward and form volcanic mountains.

Chapter 4 Study Guide

Reviewing Main Ideas

Section 1 Continental Drift

1. Alfred Wegener suggested that the continents were joined together at some point in the past in a large landmass he called Pangaea. Wegener proposed that continents have moved slowly, over millions of years, to their current locations.

2. The puzzlelike fit of the continents, fossils, climatic evidence, and similar rock structures support Wegener's idea of continental drift. However, Wegener could not explain what process could cause the movement of the landmasses.

Section 2 Seafloor Spreading

1. Detailed mapping of the ocean floor in the 1950s showed underwater mountains and rift valleys.

2. In the 1960s, Harry Hess suggested seafloor spreading as an explanation for the formation of mid-ocean ridges.

3. The theory of seafloor spreading is supported by magnetic evidence in rocks and by the ages of rocks on the ocean floor.

Section 3 Theory of Plate Tectonics

1. In the 1960s, scientists combined the ideas of continental drift and seafloor spreading to develop the theory of plate tectonics. The theory states that the surface of Earth is broken into sections called plates that move around on the asthenosphere.

2. Currents in Earth's mantle called convection currents transfer heat in Earth's interior. It is thought that this transfer of heat energy moves plates.

3. Earth is a dynamic planet. As the plates move, they interact, resulting in many of the features of Earth's surface.

Visualizing Main Ideas

Copy and complete the concept map below about continental drift, seafloor spreading, and plate tectonics.

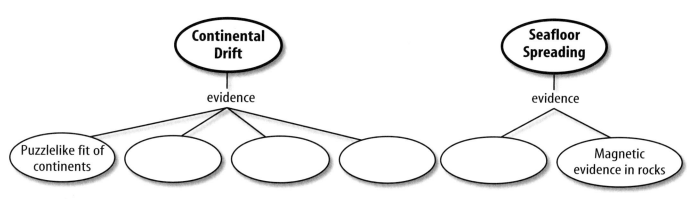

chapter 4 Review

Using Vocabulary

asthenosphere p. 106
continental drift p. 98
convection current p. 111
lithosphere p. 106
Pangaea p. 98
plate p. 106
plate tectonics p. 106
seafloor spreading p. 103

Each phrase below describes a vocabulary term from the list. Write the term that matches the phrase describing it.

1. plasticlike layer below the lithosphere
2. idea that continents move slowly across Earth's surface
3. large, ancient landmass that consisted of all the continents on Earth
4. composed of oceanic or continental crust and upper mantle
5. explains locations of mountains, trenches, and volcanoes
6. theory proposed by Harry Hess that includes processes along mid-ocean ridges

Checking Concepts

Choose the word or phrase that best answers the question.

7. Which layer of Earth contains the asthenosphere?
   A) crust
   B) mantle
   C) outer core
   D) inner core

8. What type of plate boundary is the San Andreas Fault part of?
   A) divergent
   B) subduction
   C) convergent
   D) transform

9. What hypothesis states that continents slowly moved to their present positions on Earth?
   A) subduction
   B) erosion
   C) continental drift
   D) seafloor spreading

Use the illustration below to answer question 10.

10. Which plate is subducting beneath the South American Plate?
    A) Nazca
    B) African
    C) North American
    D) Indo-Australian

11. Which of the following features are evidence that many continents were at one time near Earth's south pole?
    A) glacial deposits
    B) earthquakes
    C) volcanoes
    D) mid-ocean ridges

12. What evidence in rocks supports the theory of seafloor spreading?
    A) plate movement
    B) magnetic reversals
    C) subduction
    D) convergence

13. Which type of plate boundary is the Mid-Atlantic Ridge a part of?
    A) convergent
    B) divergent
    C) transform
    D) subduction

14. What theory states that plates move around on the asthenosphere?
    A) continental drift
    B) seafloor spreading
    C) subduction
    D) plate tectonics

120 ◆ F CHAPTER REVIEW   bookf.msscience.com/vocabulary_puzzlemaker

chapter 4 Review

Thinking Critically

15. **Infer** Why do many earthquakes but few volcanic eruptions occur in the Himalaya?

16. **Explain** Glacial deposits often form at high latitudes near the poles. Explain why glacial deposits have been found in Africa.

17. **Describe** how magnetism is used to support the theory of seafloor spreading.

18. **Explain** why volcanoes do not form along the San Andreas Fault.

19. **Explain** why the fossil of an ocean fish found on two different continents would not be good evidence of continental drift.

20. **Form Hypotheses** Mount St. Helens in the Cascade Range is a volcano. Use **Figure 9** and a U.S. map to hypothesize how it might have formed.

21. **Concept Map** Make an events-chain concept map that describes seafloor spreading along a divergent plate boundary. Choose from the following phrases: *magma cools to form new seafloor, convection currents circulate hot material along divergent boundary,* and *older seafloor is forced apart.*

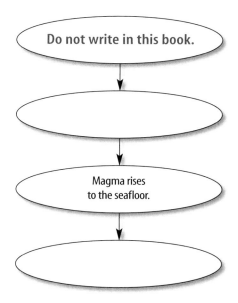

Performance Activities

22. **Observe and Infer** In the MiniLab called "Modeling Convection Currents," you observed convection currents produced in water as it was heated. Repeat the experiment, placing sequins, pieces of wood, or pieces of rubber bands into the water. How do their movements support your observations and inferences from the MiniLab?

Applying Math

23. **A Growing Rift** Movement along the African Rift Valley is about 2.1 cm per year. If plates continue to move apart at this rate, how much larger will the rift be (in meters) in 1,000 years? In 15,500 years?

Use the illustration below to answer questions 24 and 25.

24. **New Seafloor** 10 km of new seafloor has been created in 50,000 years, with 5 km on each side of a mid-ocean ridge. What is the rate of movement, in km per year, of each plate? In cm per year?

25. **Use a Ratio** If 10 km of seafloor were created in 50,000 years, how many kilometers of seafloor were created in 10,000 years? How many years will it take to create a total of 30 km of seafloor?

Chapter 4 Standardized Test Practice

Part 1 Multiple Choice

Record your answers on the answer sheet provided by your teacher or on a sheet of paper.

Use the illustration below to answer question 1.

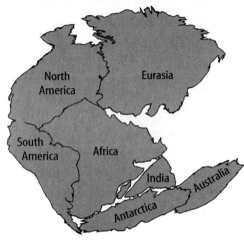

1. What is the name of the ancient supercontinent shown above?
   A. Pangaea   C. Laurasia
   B. Gondwanaland D. North America

2. Who developed the continental drift hypothesis?
   A. Harry Hess   C. Alfred Wegener
   B. J. Tuzo Wilson D. W. Jason Morgan

3. Which term refers to sections of Earth's crust and part of the upper mantle?
   A. asthenosphere   C. lithosphere
   B. plate   D. core

4. About how fast do plates move?
   A. a few millimeters each year
   B. a few centimeters each year
   C. a few meters each year
   D. a few kilometers each year

Test-Taking Tip

Marking Answers Be sure to ask if it is okay to mark in the test booklet when taking the test, but make sure you mark all answers on your answer sheet.

5. Where do Earth's plates slide past each other?
   A. convergent boundaries
   B. divergent boundaries
   C. transform boundaries
   D. subduction zones

Study the diagram below before answering questions 6 and 7.

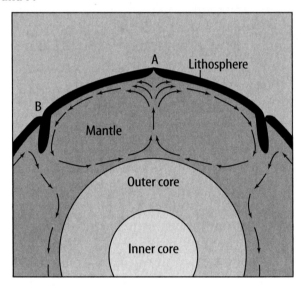

6. Suppose that the arrows in the diagram represent patterns of convection in Earth's mantle. Which type of plate boundary is most likely to occur along the region labeled "A"?
   A. transform
   B. reverse
   C. convergent
   D. divergent

7. Which statement is true of the region marked "B" on the diagram?
   A. Plates move past each other sideways.
   B. Plates move apart and volcanoes form.
   C. Plates move toward each other and volcanoes form.
   D. Plates are not moving.

Standardized Test Practice

Part 2 Short Response/Grid In

Record your answers on the answer sheet provided by your teacher or on a sheet of paper.

8. What is an ocean trench? Where do they occur?
9. How do island arcs form?
10. Why do earthquakes occur along the San Andreas Fault?
11. Describe a mid-ocean ridge.
12. Why do plates sometimes sink into the mantle?

Use the graph below to answer questions 13–15.

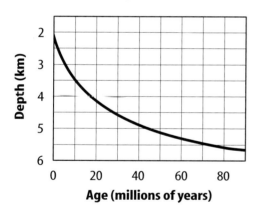

13. Use the graph to estimate the average depth below the ocean of ocean crust that has just formed.
14. Estimate the average depth of ocean crust that is 60 million years old.
15. Describe how the depth of ocean crust is related to the age of ocean crust.
16. On average, about how fast do plates move?
17. What layer in Earth's mantle do plates slide over?
18. Describe how scientists make maps of the ocean floor.

Part 3 Open Ended

Record your answers on a sheet of paper.

Use the illustration below to answer question 19.

19. Examine the diagram above. Explain how the magnetic stripes form in rock that makes up the ocean crust.
20. What causes convection in Earth's mantle?
21. Explain the theory of plate tectonics.
22. What happened to the continents that made up Pangaea after it started to break up?
23. How does Earth's lithosphere differ from Earth's asthenosphere?
24. What types of life have been discovered near mid-ocean ridges?
25. What are the three types of motion that occur at plate boundaries? Describe each motion.
26. What forms when continents collide? Describe the process.
27. What occurs at the center of a mid-ocean ridge? What might you find there?
28. What evidence do we have that supports the hypothesis of continental drift?
29. Who proposed the first theories about plate tectonics? Explain why other scientists questioned these theories.

chapter 5

Earthquakes

chapter preview

sections
1. Forces Inside Earth
2. Features of Earthquakes
   Lab Epicenter Location
3. People and Earthquakes
   Lab Earthquake Depths
• *Virtual Lab* How do seismograph stations help determine an earthquake's epicenter?

Was anyone hurt?

On October 17, 1989, the Loma Prieta earthquake rocked San Francisco, CA, leaving 62 dead and many more injured. Seismologists try to predict when and where earthquakes will occur so they can warn people of possible danger.

Science Journal Write *three* things that you would ask a scientist studying earthquakes.

Start-Up Activities

Why do earthquakes occur?

The bedrock beneath the soil can break to form cracks and move, forming faults. When blocks of bedrock move past each other along a fault, they cause the ground to shake. Why doesn't a block of bedrock move all the time, causing constant earthquakes? You'll find out during this activity.

1. Tape a sheet of medium-grain sandpaper to the tabletop.
2. Tape a second sheet of sandpaper to the cover of a textbook.
3. Place the book on the table so that both sheets of sandpaper meet.
4. Tie two large, thick rubber bands together and loop one of the rubber bands around the edge of the book so that it is not touching the sandpaper.
5. Pull on the free rubber band until the book moves. Record your observations.
6. **Think Critically** Write a paragraph that describes how the book moved. Using this model, predict why blocks of bedrock don't move all the time.

Earthquakes and Earth's Crust Make the following Foldable to help you understand the cause-and-effect relationship between earthquakes and movement in Earth's crust.

STEP 1 Fold a sheet of paper in half lengthwise.

STEP 2 Fold paper down 2.5 cm from the top. (Hint: From the tip of your index finger to your middle knuckle is about 2.5 cm.)

STEP 3 Open and draw lines along the 2.5 cm fold. Label as shown.

Read and Write As you read the chapter, write the causes and effects of earthquakes on your Foldable.

Preview this chapter's content and activities at
bookf.msscience.com

section 1

Forces Inside Earth

as you read

What You'll Learn
- **Explain** how earthquakes result from the buildup of energy in rocks.
- **Describe** how compression, tension, and shear forces make rocks move along faults.
- **Distinguish** among normal, reverse, and strike-slip faults.

Why It's Important
Earthquakes cause billions of dollars in property damage and kill an average of 10,000 people every year.

Review Vocabulary
plate: a large section of Earth's crust and rigid upper mantle that moves around on the asthenosphere

New Vocabulary
- fault
- earthquake
- normal fault
- reverse fault
- strike-slip fault

Earthquake Causes

Recall the last time you used a rubber band. Rubber bands stretch when you pull them. Because they are elastic, they return to their original shape once the force is released. However, if you stretch a rubber band too far, it will break. A wooden craft stick behaves in a similar way. When a force is first applied to the stick, it will bend and change shape. The energy needed to bend the stick is stored inside the stick as potential energy. If the force keeping the stick bent is removed, the stick will return to its original shape, and the stored energy will be released as energy of motion.

Fault Formation There is a limit to how far a wooden craft stick can bend. This is called its elastic limit. Once its elastic limit is passed, the stick remains bent or breaks, as shown in **Figure 1**. Rocks behave in a similar way. Up to a point, applied forces cause rocks to bend and stretch, undergoing what is called elastic deformation. Once the elastic limit is passed, the rocks may break. When rocks break, they move along surfaces called **faults**. A tremendous amount of force is required to overcome the strength of rocks and to cause movement along a fault. Rock along one side of a fault can move up, down, or sideways in relation to rock along the other side of the fault.

Figure 1 The bending and breaking of wooden craft sticks are similar to how rocks bend and break.

When a force is applied, the stick will bend and change shape.

When the elastic limit is passed, the stick will break.

126 ◆ F

What causes faults? What produces the forces that cause rocks to break and faults to form? The surface of Earth is in constant motion because of forces inside the planet. These forces cause sections of Earth's surface, called plates, to move. This movement puts stress on the rocks near the plate edges. To relieve this stress, the rocks tend to bend, compress, or stretch. If the force is great enough, the rocks will break. An **earthquake** is the vibrations produced by the breaking of rock. **Figure 2** shows how the locations of earthquakes outline the plates that make up Earth's surface.

Figure 2 The dots represent the epicenters of major earthquakes over a ten-year period. Note that most earthquakes occur near plate boundaries.
Form a hypothesis *to explain why earthquakes rarely occur in the middle of a plate.*

 Why do most earthquakes occur near plate boundaries?

How Earthquakes Occur As rocks move past each other along a fault, their rough surfaces catch, temporarily halting movement along the fault. However, forces keep driving the rocks to move. This action builds up stress at the points where the rocks are stuck. The stress causes the rocks to bend and change shape. When the rocks are stressed beyond their elastic limit, they can break, move along the fault, and return to their original shapes. An earthquake results. Earthquakes range from unnoticeable vibrations to devastating waves of energy. Regardless of their intensity, most earthquakes result from rocks moving over, under, or past each other along fault surfaces.

SECTION 1 Forces Inside Earth **F ◆ 127**

Figure 3 Rock above the normal fault surface moves downward in relation to rock below the fault surface. This normal fault formed near Kanab, Utah.

Figure 4 The rock above the reverse fault surface moves upward in relation to the rock below the fault surface.

Types of Faults

Three types of forces—tension, compression, and shear—act on rocks. Tension is the force that pulls rocks apart, and compression is the force that squeezes rocks together. Shear is the force that causes rocks on either side of a fault to slide past each other.

Normal Faults Tensional forces inside Earth cause rocks to be pulled apart. When rocks are stretched by these forces, a normal fault can form. Along a **normal fault,** rock above the fault surface moves downward in relation to rock below the fault surface. The motion along a normal fault is shown in **Figure 3.** Notice the normal fault shown in the photograph above.

Reverse Faults Reverse faults result from compression forces that squeeze rock. **Figure 4** shows the motion along a reverse fault. If rock breaks from forces pushing from opposite directions, rock above a **reverse fault** surface is forced up and over the rock below the fault surface. The photo below shows a large reverse fault in California.

128 ◆ **F CHAPTER 5** Earthquakes

Figure 5 Shear forces push on rock in opposite—but not directly opposite—horizontal directions. When they are strong enough, these forces split rock and create strike-slip faults.

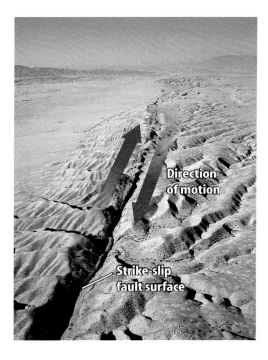

Strike-Slip Faults At a **strike-slip fault,** shown in **Figure 5,** rocks on either side of the fault are moving past each other without much upward or downward movement. The photo above shows the largest fault in California—the San Andreas Fault—which stretches more than 1,100 km through the state. The San Andreas Fault is the boundary between two of Earth's plates that are moving sideways past each other.

Reading Check *What is a strike-slip fault?*

section 1 review

Summary

Earthquake Causes
- Faults form when stressed rocks break along surfaces.
- Stresses on rock are created by plate movements.
- When rocks break along a fault, vibrations are created. This is an earthquake.

Types of Faults
- Normal faults can form when rocks undergo tension.
- Compression forces produce reverse faults.
- Strike-slip faults result when rocks move past each other without much upward or downward movement.

Self Check

1. **Infer** The Himalaya in Tibet formed when two of Earth's plates collided. What types of faults would you expect to find in these mountains? Why?
2. **State** In what direction do rocks move above a normal fault surface? What force causes this?
3. **Describe** how compression forces make rocks move along a reverse fault.
4. **Think Critically** Why is it easier to predict where an earthquake will occur than it is to predict when it will occur?

Applying Skills

5. **Infer** Why do the chances of an earthquake increase rather than decrease as time passes since the last earthquake?

section 2
Features of Earthquakes

as you read

What You'll Learn
- **Explain** how earthquake energy travels in seismic waves.
- **Distinguish** among primary, secondary, and surface waves.
- **Describe** the structure of Earth's interior.

Why It's Important
Seismic waves are responsible for most damage caused by earthquakes.

Review Vocabulary
wave: rhythmic movement that carries energy through matter and space

New Vocabulary
- seismic wave
- focus
- primary wave
- secondary wave
- surface wave
- epicenter
- seismograph

Seismic Waves

When two people hold opposite ends of a rope and shake one end, as shown in **Figure 6**, they send energy through the rope in the form of waves. Like the waves that travel through the rope, **seismic** (SIZE mihk) **waves** generated by an earthquake travel through Earth. During a strong earthquake, the ground moves forward and backward, heaves up and down, and shifts from side to side. The surface of the ground can ripple like waves do in water. Imagine trying to stand on ground that had waves traveling through it. This is what you might experience during a strong earthquake.

Origin of Seismic Waves You learned earlier that rocks move past each other along faults, creating stress at points where the rocks' irregular surfaces catch each other. The stress continues to build up until the elastic limit is exceeded and energy is released in the form of seismic waves. The point where this energy release first occurs is the **focus** (plural, *foci*) of the earthquake. The foci of most earthquakes are within 65 km of Earth's surface. A few have been recorded as deep as 700 km. Seismic waves are produced and travel outward from the earthquake focus.

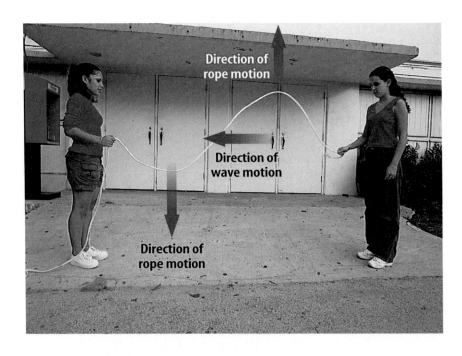

Figure 6 Some seismic waves are similar to the wave that is traveling through the rope. Note that the rope moves perpendicular to the wave direction.

Primary Waves When earthquakes occur, three different types of seismic waves are produced. All of the waves are generated at the same time, but each behaves differently within Earth. **Primary waves** (P-waves) cause particles in rocks to move back and forth in the same direction that the wave is traveling. If you squeeze one end of a coiled spring and then release it, you cause it to compress and then stretch as the wave travels through the spring, as shown in **Figure 7.** Particles in rocks also compress and then stretch apart, transmitting primary waves through the rock.

Figure 7 Primary waves move through Earth the same way that a wave travels through a coiled spring.

Secondary and Surface Waves **Secondary waves** (S-waves) move through Earth by causing particles in rocks to move at right angles to the direction of wave travel. The wave traveling through the rope shown in **Figure 6** is an example of a secondary wave.

Surface waves cause most of the destruction resulting from earthquakes. **Surface waves** move rock particles in a backward, rolling motion and a side-to-side, swaying motion, as shown in **Figure 8.** Many buildings are unable to withstand intense shaking because they are made with stiff materials. The buildings fall apart when surface waves cause different parts of the building to move in different directions.

Sound Waves When sound is produced, waves move through air or some other material. Research sound waves to find out which type of seismic wave they are similar to.

Reading Check *Why do surface waves damage buildings?*

Surface waves are produced when earthquake energy reaches the surface of Earth. Surface waves travel outward from the epicenter. The earthquake **epicenter** (EH pih sen tur) is the point on Earth's surface directly above the earthquake focus. Find the focus and epicenter in **Figure 9.**

Figure 8 Surface waves move rock particles in a backward, rolling motion and a side-to-side, swaying motion.
Compare and contrast *surface waves and secondary waves.*

NATIONAL GEOGRAPHIC VISUALIZING SEISMIC WAVES

Figure 9

As the plates that form Earth's lithosphere move, great stress is placed on rocks. They bend, stretch, and compress. Occasionally, rocks break, producing earthquakes that generate seismic waves. As shown here, different kinds of seismic waves—each with distinctive characteristics—move outward from the focus of the earthquake.

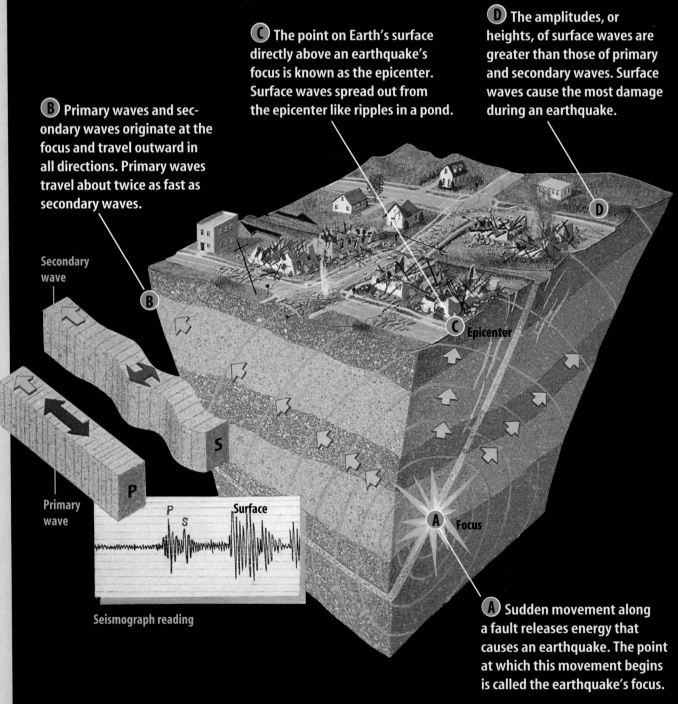

B Primary waves and secondary waves originate at the focus and travel outward in all directions. Primary waves travel about twice as fast as secondary waves.

C The point on Earth's surface directly above an earthquake's focus is known as the epicenter. Surface waves spread out from the epicenter like ripples in a pond.

D The amplitudes, or heights, of surface waves are greater than those of primary and secondary waves. Surface waves cause the most damage during an earthquake.

A Sudden movement along a fault releases energy that causes an earthquake. The point at which this movement begins is called the earthquake's focus.

132 ◆ F CHAPTER 5 Earthquakes

Locating an Epicenter

Different seismic waves travel through Earth at different speeds. Primary waves are the fastest, secondary waves are slower, and surface waves are the slowest. Can you think of a way this information could be used to determine how far away an earthquake epicenter is? Think of the last time you saw two people running in a race. You probably noticed that the faster person got further ahead as the race continued. Like runners in a race, seismic waves travel at different speeds.

Scientists have learned how to use the different speeds of seismic waves to determine the distance to an earthquake epicenter. When an epicenter is far from a location, the primary wave has more time to put distance between it and the secondary and surface waves, just like the fastest runner in a race.

Measuring Seismic Waves Seismic waves from earthquakes are measured with an instrument known as a **seismograph.** Seismographs register the waves and record the time that each arrived. Seismographs consist of a rotating drum of paper and a pendulum with an attached pen. When seismic waves reach the seismograph, the drum vibrates but the pendulum remains at rest. The stationary pen traces a record of the vibrations on the moving drum of paper. The paper record of the seismic event is called a seismogram. **Figure 10** shows two types of seismographs that measure either vertical or horizontal ground movement, depending on the orientation of the drum.

Science online

Topic: Earthquake Data
Visit bookf.msscience.com for Web links to the National Earthquake Information Center and the World Data Center for Seismology.

Activity List the locations and distances of each reference that seigmograph stations used to determine the epicenter of the most recent earthquake.

Figure 10 Seismographs differ according to whether they are intended to measure horizontal or vertical seismic motions.
Infer why one seismograph can't measure both horizontal and vertical motions.

SECTION 2 Features of Earthquakes **F ◆ 133**

Seismograph Stations Each type of seismic wave reaches a seismograph station at a different time based on its speed. Primary waves arrive first at seismograph stations, and secondary waves, which travel slower, arrive second. Because surface waves travel slowest, they arrive at seismograph stations last. This difference in arrival times is used to calculate the distance from the seismograph station to the earthquake epicenter, as shown in **Figure 11**. If a seismograph station is located 4,000 km from an earthquake epicenter, primary waves will reach the station about 6 minutes before secondary waves.

If seismic waves reach three or more seismograph stations, the location of the epicenter can be determined. To locate an epicenter, scientists draw circles around each station on a map. The radius of each circle equals that station's distance from the earthquake epicenter. The point where all three circles intersect, shown in **Figure 12,** is the location of the earthquake epicenter.

Seismologists usually describe earthquakes based on their distances from the seismograph. Local events occur less than 100 km away. Regional events occur 100 km to 1,400 km away. Teleseismic events are those that occur at distances greater than 1,400 km.

Figure 11 Primary waves arrive at a seismograph station before secondary waves do.
Use Graphs *If primary waves reach a seismograph station two minutes before secondary waves, how far is the station from the epicenter?*

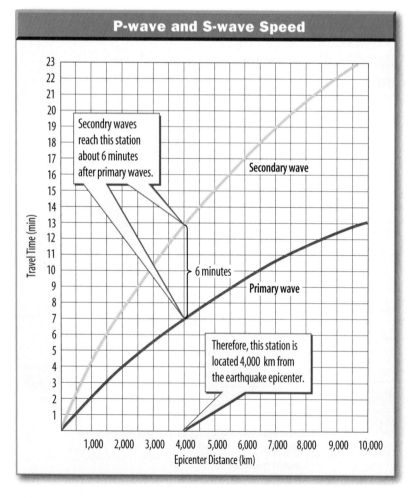

Figure 12 The radius of each circle is equal to the distance from the epicenter to each seismograph station. The intersection of the three circles is the location of the epicenter.

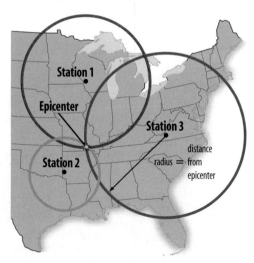

Basic Structure of Earth

Figure 13 shows Earth's internal structure. At the very center of Earth is a solid, dense inner core made mostly of iron with smaller amounts of nickel, oxygen, silicon, and sulfur. Pressure from the layers above causes the inner core to be solid. Above the solid inner core lies the liquid outer core, which also is made mainly of iron.

Reading Check *How do the inner and outer cores differ?*

Earth's mantle is the largest layer, lying directly above the outer core. It is made mostly of silicon, oxygen, magnesium, and iron. The mantle often is divided into an upper part and a lower part based on changing seismic wave speeds. A portion of the upper mantle, called the asthenosphere (as THE nuh sfihr), consists of weak rock that can flow slowly.

Earth's Crust The outermost layer of Earth is the crust. Together, the crust and a part of the mantle just beneath it make up Earth's lithosphere (LIH thuh sfihr). The lithosphere is broken into a number of plates that move over the asthenosphere beneath it.

The thickness of Earth's crust varies. It is more than 60 km thick in some mountainous regions and less than 5 km thick under some parts of the oceans. Compared to the mantle, the crust contains more silicon and aluminum and less magnesium and iron. Earth's crust generally is less dense than the mantle beneath it.

Mini LAB

Interpreting Seismic Wave Data

Procedure
Copy the table below into your **Science Journal**. Use the **graph** in **Figure 11** to determine the difference in arrival times for primary and secondary waves at the distances listed in the data table below. Two examples are provided for you.

Wave Data	
Distance (km)	Difference in Arrival Time
1,500	2 min, 50 s
2,250	Do not write in this book.
2,750	
3,000	
4,000	5 min, 55 s
7,000	
9,000	

Analysis
1. What happens to the difference in arrival times as the distance from the earthquake increases?
2. If the difference in arrival times at a seismograph station is 6 min, 30 s, how far away is the epicenter?

Figure 13 The internal structure of Earth shows that it is made of different layers.

Mapping Earth's Internal Structure As shown in **Figure 14,** the speeds and paths of seismic waves change as they travel through materials with different densities. By studying seismic waves that have traveled through Earth, scientists have identified different layers with different densities. In general, the densities increase with depth as pressures increase. Studying seismic waves has allowed scientists to map Earth's internal structure without being there.

Early in the twentieth century, scientists discovered that large areas of Earth don't receive seismic waves from an earthquake. In the area on Earth between 105° and 140° from the earthquake focus, no waves are detected. This area, called the shadow zone, is shown in **Figure 14.** Secondary waves are not transmitted through a liquid, so they stop when they hit the liquid outer core. Primary waves are slowed and bent but not stopped by the liquid outer core. Because of this, scientists concluded that the outer core and mantle are made of different materials. Primary waves speed up again as they travel through the solid inner core. The bending of primary waves and the stopping of secondary waves create the shadow zone.

Reading Check *Why do seismic waves change speed as they travel through Earth?*

Figure 14 Seismic waves bend and change speed as the density of rock changes. Primary waves bend when they contact the outer core, and secondary waves are stopped completely. This creates a shadow zone where no seismic waves are received.

Layer Boundaries Figure 15 shows how seismic waves change speed as they pass through layers of Earth. Seismic waves speed up when they pass through the bottom of the crust and enter the upper mantle, shown on the far left of the graph. This boundary between the crust and upper mantle is called the Mohorovicic discontinuity (moh huh ROH vee chihch • dis kahn tuh NEW uh tee), or Moho.

The mantle is divided into layers based on changes in seismic wave speeds. For example, primary and secondary waves slow down again when they reach the asthenosphere. Then they generally speed up as they move through a more solid region of the mantle below the asthenosphere.

The core is divided into two layers based on how seismic waves travel through it. Secondary waves do not travel through the liquid core, as you can see in the graph. Primary waves slow down when they reach the outer core, but they speed up again upon reaching the solid inner core.

Figure 15 Changes in the speeds of seismic waves allowed scientists to detect boundaries between Earth's layers. S waves in the inner core form when P waves strike its surface.

section 2 review

Summary

Seismic Waves
- Stress builds up at the points where the surfaces of the fault touch.
- When the elastic limit of the rock is exceeded, it moves, producing seismic waves.
- There are three types of seismic waves—primary, secondary, and surface.

Locating an Epicenter
- A seismograph measures seismic waves.
- Three seismograph stations are needed to determine the location of an epicenter.

Basic Structure of Earth
- The inner core, the outer core, the lower mantle, the upper mantle, and the crust make up Earth.

Self Check

1. **Compare and contrast** the movement of rocks by primary waves, secondary waves, and surface waves.
2. **Explain** why surface waves cause the most damage to property.
3. **Describe** what makes up most of Earth's inner core.
4. **Explain** why three seismograph stations are needed to determine the location of an epicenter.
5. **Think Critically** Why do some seismograph stations receive both primary and secondary waves from an earthquake but other stations don't?

Applying Skills

6. **Simple Equations** Primary waves travel about 6 km/s through Earth's crust. The distance from Los Angeles to Phoenix is about 600 km. How long would it take primary waves to travel between the two cities?

Epicenter Location

In this lab you can plot the distance of seismograph stations from the epicenters of earthquakes and determine the location of earthquake epicenters.

Earthquake Data

Location of Seismograph	Wave	Wave Arrival Times Earthquake A	Wave Arrival Times Earthquake B
New York, New York	P	2:24:05 P.M.	1:19:42 P.M.
	S	2:29:15 P.M.	1:25:27 P.M.
Seattle, Washington	P	2:24:40 P.M.	1:14:37 P.M.
	S	2:30:10 P.M.	1:16:57 P.M.
Rio de Janeiro, Brazil	P	2:29:10 P.M.	—
	S	2:37:50 P.M.	—
Paris, France	P	2:30:30 P.M.	1:24:57 P.M.
	S	2:40:10 P.M.	1:34:27 P.M.
Tokyo, Japan	P	—	1:24:27 P.M.
	S	—	1:33:27 P.M.

▶ Real-World Question

How can plotting the distance of several seismograph stations from an earthquake epicenter allow you to determine the locations of the epicenter?

Goals

- **Plot** the distances from several seismograph stations based on primary and secondary wave arrival times.
- **Interpret** the location of earthquake epicenters from these plots.

Materials

string   globe
metric ruler chalk

▶ Procedure

1. Determine the difference in arrival time between the primary and secondary waves at each station for each earthquake listed in the table.

2. After you determine the arrival time differences for each seismograph station, use the graph in **Figure 11** to determine the distance in kilometers of each seismograph from the epicenter of each earthquake. Record these data in a data table. For example, the difference in arrival times in Paris for earthquake B is 9 min, 30 s. On the graph, the primary and secondary waves are separated along the vertical axis by 9 min, 30 s at a distance of 8,975 km.

3. Using the string, measure the circumference of the globe. Determine a scale of centimeters of string to kilometers on Earth's surface. (Earth's circumference is 40,000 km.)

4. For each earthquake, place one end of the string at each seismic station location on the globe. Use the chalk to draw a circle with a radius equal to the distance to the earthquake's epicenter.

5. **Identify** the epicenter for each earthquake.

▶ Conclude and Apply

1. How is the distance of a seismograph from the earthquake related to the arrival times of the waves?
2. **Identify** the location of the epicenter for each earthquake.
3. How many stations were needed to locate each epicenter accurately?
4. **Explain** why some seismographs didn't receive seismic waves from some quakes.

section 3
People and Earthquakes

Earthquake Activity

Imagine waking up in the middle of the night with your bed shaking, windows shattering, and furniture crashing together. That's what many people in Northridge, California, experienced at 4:30 A.M. on January 17, 1994. The ground beneath Northridge shook violently—it was an earthquake.

Although the earthquake lasted only 15 s, it killed 51 people, injured more than 9,000 people, and caused $44 billion in damage. More than 22,000 people were left homeless. **Figure 16** shows some of the damage caused by the Northridge earthquake and a seismogram made by that quake.

Earthquakes are natural geological events that provide information about Earth. Unfortunately, they also cause billions of dollars in property damage and kill an average of 10,000 people every year. With so many lives lost and such destruction, it is important for scientists to learn as much as possible about earthquakes to try to reduce their impact on society.

Figure 16 The 1994 Northridge, California, earthquake was a costly disaster. Several major highways were damaged and 51 lives were lost.

as you read

What You'll Learn
- **Explain** where most earthquakes in the United States occur.
- **Describe** how scientists measure earthquakes.
- **List** ways to make your classroom and home more earthquake-safe.

Why It's Important
Earthquake preparation can save lives and reduce damage.

Review Vocabulary
crest: the highest point of a wave

New Vocabulary
- magnitude
- liquefaction
- tsunami

Seismogram of the Northridge earthquake

Studying Earthquakes Scientists who study earthquakes and seismic waves are seismologists. As you learned earlier, the instrument that is used to record primary, secondary, and surface waves from earthquakes all over the world is called a seismograph. Seismologists can use records from seismographs, called seismograms, to learn more than just where the epicenter of an earthquake is located.

Measuring Earthquake Magnitude The height of the lines traced on the paper of a seismograph is a measure of the energy that is released, or the **magnitude,** of the earthquake. The Richter magnitude scale is used to describe the strength of an earthquake and is based on the height of the lines on the seismogram. The Richter scale has no upper limit. However, scientists think that a value of about 9.5 would be the maximum strength an earthquake could register. For each increase of 1.0 on the Richter scale, the height of the line on a seismogram is ten times greater. However, about 32 times as much energy is released for every increase of 1.0 on the scale. For example, an earthquake with a magnitude of 8.5 releases about 32 times more energy than an earthquake with a magnitude of 7.5.

Figure 17 The 1999 earthquake in Turkey released about 32 times more energy than the 1994 Northridge earthquake did.

Past Earthquakes Damage from the 7.8-magnitude earthquake in Turkey in 1999 is shown in **Figure 17. Table 1** is a list of some large-magnitude earthquakes that have occurred around the world and the damage they have caused. Most of the earthquakes you hear about are large ones that cause great damage. However, of all the earthquakes detected throughout the world each year, most have magnitudes too low to be felt by humans. Scientists record thousands of earthquakes every day with magnitudes of less than 3.0. Each year, about 55,000 earthquakes are felt but cause little or no damage. These minor earthquakes have magnitudes that range from approximately 3.0 to 4.9 on the Richter scale.

Table 1	Large-Magnitude Earthquakes		
Year	Location	Magnitude	Deaths
1556	Shensi, China	?	830,000
1886	Charleston, SC	?	60
1906	San Francisco, CA	8.3	700 to 800
1923	Tokyo, Japan	9.2	143,000
1960	Chile	9.5	490 to 2,290
1975	Laoning Province, China	7.5	few
1976	Tangshan, China	8.2	242,000
1990	Iran	7.7	50,000
1994	Northridge, CA	6.8	51
2001	India	7.7	>20,000
2003	Northern Algeria	6.8	2,266

Describing Earthquake Intensity Earthquakes also can be described by the amount of damage they cause. The modified Mercalli intensity scale describes the intensity of an earthquake using the amount of structural and geologic damage in a specific location. The amount of damage done depends on the strength of the earthquake, the nature of surface material, the design of structures, and the distance from the epicenter.

Under ideal conditions, only a few people would feel an intensity-I earthquake, and it would cause no damage. An intensity-IV earthquake would be felt by everyone indoors during the day but would be felt by only a few people outdoors. Pictures might fall off walls and books might fall from shelves. However, an intensity-IX earthquake would cause considerable damage to buildings and would cause cracks in the ground. An intensity-XII earthquake would cause total destruction of buildings, and objects such as cars would be thrown upward into the air. The 1994 6.8-magnitude earthquake in Northridge, California, was listed at an intensity of IX because of the damage it caused.

Liquefaction Have you ever tried to drink a thick milkshake from a cup? Sometimes the milkshake is so thick that it won't flow. How do you make the milkshake flow? You shake it. Something similar can happen to very wet soil during an earthquake. Wet soil can be strong most of the time, but the shaking from an earthquake can cause it to act more like a liquid. This is called **liquefaction.** When liquefaction occurs in soil under buildings, the buildings can sink into the soil and collapse, as shown in **Figure 18.** People living in earthquake regions should avoid building on loose soils.

Magnetism In 1975, Chinese scientists successfully predicted an earthquake by measuring a slow tilt of Earth's surface and small changes in Earth's magnetism. Many lives were saved as a result of this prediction. Research the jobs that seismologists do and the types of organizations that they work for. Find out why most earthquakes have not been predicted.

Figure 18 San Francisco's Marina district suffered extensive damage from liquefaction in the 1989 Loma Prieta earthquake because it is built on a landfilled marsh.

Topic: Tsunamis
Visit bookf.msscience.com for Web links to information about tsunamis.

Activity Make a table that displays the location, date, earthquake magnitude, and maximum runup of the five most recent tsunamis.

Tsunamis Most earthquake damage occurs when surface waves cause buildings, bridges, and roads to collapse. People living near the seashore, however, have another problem. An earthquake under the ocean causes a sudden movement of the ocean floor. The movement pushes against the water, causing a powerful wave that can travel thousands of kilometers in all directions.

Ocean waves caused by earthquakes are called seismic sea waves, or **tsunamis** (soo NAH meez). Far from shore, a wave caused by an earthquake is so long that a large ship might ride over it without anyone noticing. But when one of these waves breaks on a shore, as shown in **Figure 19,** it forms a towering crest that can reach 30 m in height.

Tsunami Warnings Just before a tsunami crashes onto shore, the water along a shoreline might move rapidly toward the sea, exposing a large portion of land that normally is underwater. This should be taken as a warning sign that a tsunami could strike soon. You should head for higher ground immediately.

Because of the number of earthquakes that occur around the Pacific Ocean, the threat of tsunamis is constant. To protect lives and property, a warning system has been set up in coastal areas and for the Pacific Islands to alert people if a tsunami is likely to occur. The Pacific Tsunami Warning Center, located near Hilo, Hawaii, provides warning information including predicted tsunami arrival times at coastal areas.

However, even tsunami warnings can't prevent all loss of life. In the 1960 tsunami that struck Hawaii, 61 people died when they ignored the warning to move away from coastal areas.

Figure 19 A tsunami begins over the earthquake focus.
Infer what might happen to towns located near the shore.

Earthquake Safety

Although earthquakes cannot be predicted reliably, **Figure 20** shows where earthquakes are most likely to occur in the United States. Knowing where earthquakes are likely to occur helps in long-term planning. Cities can take action to reduce damage and loss of life. Many buildings withstood the 1989 Loma Prieta earthquake because they were built with the expectation that such an earthquake would occur someday.

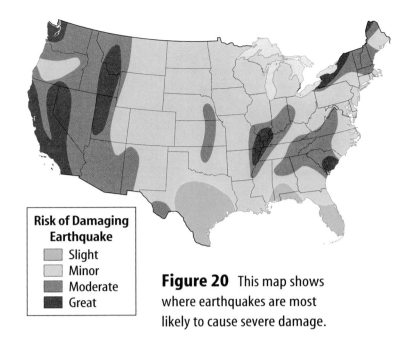

Risk of Damaging Earthquake
- Slight
- Minor
- Moderate
- Great

Figure 20 This map shows where earthquakes are most likely to cause severe damage.

Applying Math — Find a Ratio

EARTHQUAKE ENERGY An increase of one magnitude on the Richter scale for an earthquake means that 32 times more energy is released. How many times greater is the energy released by a magnitude-6 earthquake than the energy released by a magnitude-3 earthquake?

Solution

1 *This is what you know:*
- magnitude-6 earthquake, magnitude-3 earthquake
- energy increases 32 times per magnitude number

2 *This is what you need to find out:* How many times greater is the energy of the larger earthquake than the energy of the smaller earthquake?

3 *This is the procedure you need to use:*
- Find the difference in magnitudes: $6 - 3 = 3$.
- This is the number of times 32 is multiplied times itself. $32 \times 32 \times 32 = 32{,}768$.
- The magnitude-6 earthquake releases 32,768 times more energy than the magnitude-3 earthquake.

4 *Check your answer:* Count how many times you need to divide 32,768 by 32 to obtain 1. You should get 3.

Practice Problems

1. How many times greater is the energy released by a magnitude-7 earthquake than the energy released by a magnitude-2 earthquake?

2. How many times greater is the energy released by a magnitude-5 earthquake than the energy released by a magnitude-3 earthquake?

For more practice, visit bookf.msscience.com/math_practice

Mini LAB

Modeling Seismic-Safe Structures

Procedure

1. On a **tabletop,** build a structure out of **building blocks** by simply placing one block on top of another.
2. Build a second structure by wrapping sections of three blocks together with **rubber bands.** Then, wrap larger rubber bands around the entire completed structure.
3. Set the second structure on the tabletop next to the first one and pound on the side of the table with a slow, steady rhythm.

Analysis

1. Which of your two structures was better able to withstand the "earthquake" caused by pounding on the table?
2. How might the idea of wrapping the blocks with rubber bands be used in construction of supports for elevated highways?

Try at Home

Quake-Resistant Structures During earthquakes, buildings, bridges, and highways can be damaged or destroyed. Most loss of life during an earthquake occurs when people are trapped in or on these crumbling structures. What can be done to reduce loss of life?

Seismic-safe structures stand up to vibrations that occur during an earthquake. **Figure 21** shows how buildings can be built to resist earthquake damage. Today in California, some new buildings are supported by flexible, circular moorings placed under the buildings. The moorings are made of steel plates filled with alternating layers of rubber and steel. The rubber acts like a cushion to absorb earthquake waves. Tests have shown that buildings supported in this way should be able to withstand an earthquake measuring up to 8.3 on the Richter scale without major damage.

In older buildings, workers often install steel rods to reinforce building walls. Such measures protect buildings in areas that are likely to experience earthquakes.

Reading Check *What are seismic-safe structures?*

Figure 21 The rubber portions of this building's moorings absorb most of the wave motion of an earthquake. The building itself only sways gently.
Infer *what purpose the rubber serves.*

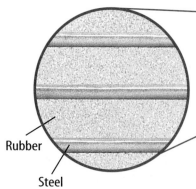

Rubber

Steel

144 ◆ **F CHAPTER 5** Earthquakes

Before an Earthquake To make your home as earthquake-safe as possible, certain steps can be taken. To reduce the danger of injuries from falling objects, move heavy objects from high shelves to lower shelves. Learn how to turn off the gas, water, and electricity in your home. To reduce the chance of fire from broken gas lines, make sure that water heaters and other gas appliances are held securely in place as shown in **Figure 22.** A newer method that is being used to minimize the danger of fire involves placing sensors on gas lines. The sensors automatically shut off the gas when earthquake vibrations are detected.

During an Earthquake If you're indoors, move away from windows and any objects that could fall on you. Seek shelter in a doorway or under a sturdy table or desk. If you're outdoors, stay in the open—away from power lines or anything that might fall. Stay away from chimneys or other parts of buildings that could fall on you.

After an Earthquake If water and gas lines are damaged, the valves should be shut off by an adult. If you smell gas, leave the building immediately and call authorities from a phone away from the leak area. Stay away from damaged buildings. Be careful around broken glass and rubble, and wear boots or sturdy shoes to keep from cutting your feet. Finally, stay away from beaches. Tsunamis sometimes hit after the ground has stopped shaking.

Figure 22 Sturdy metal straps on this gas water heater help reduce the danger of fires from broken gas lines during an earthquake.

section 3 review

Summary

Earthquake Activity

- The height of the lines traced on a seismogram can be used to determine an earthquake's magnitude.
- The intensity of an earthquake is determined by examining the amount of damage caused by the earthquake.

Earthquake Safety

- Knowing where large earthquakes are likely to occur helps people plan how to reduce damage.
- If you're ever in an earthquake, move away from windows or any object that might fall on you. Seek shelter in a doorway or under a sturdy table or desk.

Self Check

1. **Explain** how you can determine if you live in an area where an earthquake is likely to occur.
2. **Compare and contrast** the Richter and the Mercalli scales.
3. **Explain** what causes a tsunami.
4. **Describe** three ways an earthquake causes damage.
5. **Think Critically** How are shock absorbers on a car similar to the circular moorings used in modern earthquake-safe buildings? How do they absorb shock?

Applying Skills

6. **Infer** Seismographs around the world record the occurrence of thousands of earthquakes every day. Why are so few earthquakes in the news?

Earthquake Depths

Goals
- **Observe** any connection between earthquake-focus depth and epicenter location using the data provided on the next page.
- **Describe** any observed relationship between earthquake-focus depth and the movement of plates at Earth's surface.

Materials
graph paper
pencil

Real-World Question

You learned in this chapter that Earth's crust is broken into sections called plates. Stresses caused by movement of plates generate energy within rocks that must be released. When this release is sudden and rocks break, an earthquake occurs. Can a study of the foci of earthquakes tell you about plate movement in a particular region?

Analyze Your Data

1. Use graph paper and the data table on the right to make a graph plotting the depths of earthquake foci and the distances from the coast of a continent for each earthquake epicenter.

2. Use the graph below as a reference to draw your own graph. Place *Distance from the coast* and units on the *x*-axis. Begin labeling at the far left with 100 km west. To the right of it should be 0 km, then 100 km east, 200 km east, 300 km east, and so on through 700 km east. What point on your graph represents the coast?

3. Label the *y*-axis *Depth below Earth's surface.* Label the top of the graph *0 km* to represent Earth's surface. Label the bottom of the *y*-axis *−800 km*.

4. **Plot** the focus depths against the distance and direction from the coast for each earthquake in the table below.

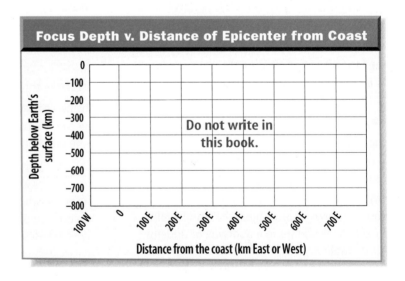

146 ◆ F **CHAPTER 5** Earthquakes

Using Scientific Methods

Conclude and Apply

1. **Describe** any observed relationship between the location of earthquake epicenters and the depth of foci.
2. **Explain** why none of the plotted earthquakes occurred below 700 km.
3. Based on your graph, form a hypothesis to explain what is happening to the plates at Earth's surface in the vicinity of the plotted earthquake foci. In what direction are the plates moving relative to each other?
4. **Infer** what process is causing the earthquakes you plotted on your graph.
5. **Infer** whether these earthquakes are occurring along the eastern side of a continent or along the western side of a continent.
6. **Draw and label** a cross section of the Earth beneath this coast. Label the eastern plate, the western plate, and use arrows to show the directions the plates are moving.
7. **Form a hypothesis** to predict which continent these data might apply to. Apply what you have learned in this lab and the information in **Figure 2**. Explain your answer.

Focus and Epicenter Data

Earthquake	Focus Depth (km)	Distance of Epicenter from Coast (km)
A	−55	0
B	−295	100 east
C	−390	455 east
D	−60	75 east
E	−130	255 east
F	−195	65 east
G	−695	400 east
H	−20	40 west
I	−505	695 east
J	−520	390 east
K	−385	335 east
L	−45	95 east
M	−305	495 east
N	−480	285 east
O	−665	545 east
P	−85	90 west
Q	−525	205 east
R	−85	25 west
S	−445	595 east
T	−635	665 east
U	−55	95 west
V	−70	100 west

Communicating Your Data

Compare your graph with those of other members of your class. **For more help, refer to the** Science Skill Handbook.

SCIENCE Stats

Moving Earth!

Did you know...

... Tsunamis can travel as fast as commercial jets and can reach heights of 30 m. A wave that tall would knock over this lighthouse. Since 1945, more people have been killed by tsunamis than by the ground shaking from earthquakes.

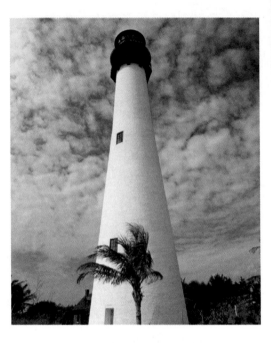

... The most powerful earthquake to hit the United States in recorded history shook Alaska in 1964. At 8.5 on the Richter scale, the quake shook all of Alaska for nearly 5 min, which is a long time for an earthquake. Nearly 320 km of roads near Anchorage suffered damage, and almost half of the 204 bridges had to be rebuilt.

Applying Math How many 3.0-magnitude earthquakes would it take to equal the energy released by one 8.0-magnitude earthquake?

... Snakes can sense the vibrations made by a small rodent up to 23 m away. Does this mean that they can detect vibrations prior to major earthquakes? Unusual animal behavior was observed just before a 1969 earthquake in China— an event that was successfully predicted.

Write About It

Visit bookf.msscience.com/science_stats to research the history and effects of earthquakes in the United States. In a paragraph, describe how the San Francisco earthquake of 1906 affected earthquake research.

chapter 5 Study Guide

Reviewing Main Ideas

Section 1 — Forces Inside Earth

1. Plate movements can cause rocks to bend and stretch. Rocks can break if the forces on them are beyond their elastic limit.
2. Earthquakes are vibrations produced when rocks break along a fault.
3. Normal faults form when rocks are under tension. Reverse faults form under compression and shearing forces produce strike-slip faults.

Section 2 — Features of Earthquakes

1. Primary waves stretch and compress rock particles. Secondary waves move particles at right angles to the direction of wave travel.
2. Surface waves move rock particles in a backward, rolling motion and a side-to-side swaying motion.
3. Earthquake epicenters are located by recording seismic waves.
4. The boundaries between Earth's internal layers are determined by observing the speeds and paths of seismic waves.

Section 3 — People and Earthquakes

1. A seismograph measures the magnitude of an earthquake.
2. The magnitude of an earthquake is related to the energy released by the earthquake.

Visualizing Main Ideas

Copy and complete the following concept map on earthquake damage.

bookf.msscience.com/interactive_tutor

chapter 5 Review

Using Vocabulary

earthquake p. 127
epicenter p. 131
fault p. 126
focus p. 130
liquefaction p. 141
magnitude p. 140
normal fault p. 128
primary wave p. 131
reverse fault p. 128
secondary wave p. 131
seismic wave p. 130
seismograph p. 133
strike-slip fault p. 129
surface wave p. 131
tsunami p. 142

Fill in the blanks with the correct words.

1. _____ causes most of the damage in earthquakes because of the side to side swaying motion that many buildings are unable to withstand.

2. At a(n) _____, rocks move past each other without much upward or downward movement.

3. The point on Earth's surface directly above the earthquake focus is the _____.

4. The measure of the energy released during an earthquake is its _____.

5. An earthquake under the ocean can cause a(n) _____ that travels thousands of kilometers.

Checking Concepts

Choose the word or phrase that best answers the question.

6. Earthquakes can occur when which of the following is passed?
   A) tension limit   C) elastic limit
   B) seismic unit   D) shear limit

7. When the rock above the fault surface moves down relative to the rock below the fault surface, what kind of fault forms?
   A) normal   C) reverse
   B) strike-slip D) shear

Use the illustration below to answer question 8.

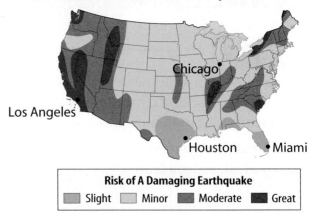

8. Using the figure above, which city should be most prepared for an earthquake?
   A) Miami, FL   C) Chicago, IL
   B) Houston, TX   D) Los Angeles, CA

9. From which of the following do primary and secondary waves move outward?
   A) epicenter   C) Moho
   B) focus   D) tsunami

10. What kind of earthquake waves stretch and compress rocks?
    A) surface   C) secondary
    B) primary   D) shear

11. What are the slowest seismic waves?
    A) surface   C) secondary
    B) primary   D) pressure

12. What is the fewest number of seismograph stations that are needed to locate the epicenter of an earthquake?
    A) two   C) four
    B) three   D) five

13. What happens to primary waves when they pass from liquids into solids?
    A) slow down   C) stay the same
    B) speed up   D) stop

14. What part of a seismograph does not move during an earthquake?
    A) sheet of paper   C) drum
    B) fixed frame   D) pendulum

150 ◆ F CHAPTER REVIEW

chapter 5 Review

Thinking Critically

15. **Infer** The 1960 earthquake in the Pacific Ocean off the coast of Chile caused damage and loss of life in Chile, Hawaii, Japan, and other areas along the Pacific Ocean border. How could this earthquake do so much damage to areas thousands of kilometers from its epicenter?

16. **Explain** why a person who is standing outside in an open field is relatively safe during a strong earthquake.

17. **Describe** how a part of the seismograph remains at rest during an earthquake.

18. **Explain** why it is incorrect to call a tsunami a tidal wave.

19. **Predict** which is likely to be more stable during an earthquake—a single-story wood-frame house or a brick building. Explain.

20. **Measure in SI** Use an atlas and a metric ruler to answer the following question. Primary waves travel at about 6 km/s in continental crust. How long would it take a primary wave to travel from San Francisco, California, to Reno, Nevada?

Use the table below to answer question 21.

Seismograph Station Data			
Station	Latitude	Longitude	Distance from Earthquake
1	45° N	120° W	1,300 km
2	35° N	105° W	1,200 km
3	40° N	115° W	790 km

21. **Use Tables** Use a map of the United States that has a distance scale, a compass for drawing circles, and the table above to determine the location of the earthquake epicenter.

Performance Activities

22. **Model** Use layers of different colors of clay to illustrate the three different kinds of faults. Label each model, explaining the forces involved and the rock movement.

Applying Math

23. **Earthquake Magnitude** An increase of one on the Richter scale corresponds to an increase of 10 in the size of the largest wave on a seismogram. How many times larger is the largest wave of a Richter magnitude-6 earthquake than a Richter magnitude-3 earthquake?

24. **Tsunami Speed** An underwater earthquake produces a tsunami 1,500 km away from Hawaii. If the tsunami travels at 600 km/h, how long will it take to reach Hawaii?

Use the graph below to answer question 25.

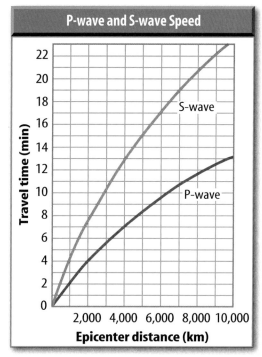

25. **Earthquake Waves** The graph above shows a P-wave and an S-wave plotted on a time-distance graph. According to the graph, which wave moves at the greater speed?

chapter 5 Standardized Test Practice

Part 1 Multiple Choice

Record your answers on the answer sheet provided by your teacher or on a sheet of paper.

Use the photo below to answer question 1.

1. The instrument above records seismic waves from an earthquake. Which of the following is the name of this instrument?
   A. seismogram   C. tiltmeter
   B. seismograph D. strainmeter

2. Which of the following terms is used to indicate the region where no earthquake waves reach Earth's surface?
   A. light zone   C. shadow zone
   B. waveless zone D. seismic zone

3. Which is used to measure magnitude?
   A. Richter scale   C. shadow zone
   B. Mercalli scale D. seismic gap

4. What is earthquake intensity?
   A. a measure of energy released
   B. a measure of seismic risk
   C. a measure of damage done
   D. a measure of an earthquake's focus

> **Test-Taking Tip**
>
> **Eliminate Incorrect Answers** If you don't know the answer to a multiple choice question, try to eliminate as many incorrect answers as possible.

5. Which of the following describes liquefaction?
   A. the stopping of S-waves by Earth's molten outer core
   B. ice melting during an earthquake to cause flooding
   C. seismic waves shaking sediment, causing it to become more liquid like
   D. rivers diverted by the motion of earthquake flooding

6. Which of the following describes the motion of secondary waves?
   A. a backward rolling and a side-to-side swaying motion
   B. a back-and-forth motion that is parallel to the direction of travel
   C. vibration in directions that are perpendicular to the direction of wave travel
   D. a forward rolling motion

Use the illustration below to answer question 7.

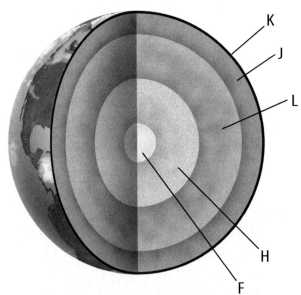

7. Which letter corresponds to the lower mantle?
   A. F   C. L
   B. H   D. J

152 ◆ F STANDARDIZED TEST PRACTICE

Standardized Test Practice

Part 2 Short Response/Grid In

Record your answers on the answer sheet provided by your teacher or on a sheet of paper.

8. Explain how earthquakes occur. Include a description of how energy builds up in rocks and is later released.

Use the illustration below to answer questions 9–10.

9. What type of fault is shown?
10. What type of force caused this fault to form?
11. Where is the fault plane?
12. Explain why most earthquakes occur along plate boundaries.
13. An earthquake that occurred in San Fernando, California, in 1971 caused $500 million in damage. An earthquake that occurred in Whittier, California, in 1987 caused $358 million in damage. How much more damage was caused by the San Fernando earthquake?
14. What factors affect damage done by an earthquake?
15. Explain the relationship between worldwide earthquake distribution and tectonic boundaries.

Part 3 Open Ended

Record your answers on a sheet of paper.

16. The three types of faults are normal faults, reverse faults, and strike-slip faults. Draw each type of fault including arrows that show which way the rocks move.
17. Describe how a person should prepare for an earthquake and how a person should react if an earthquake occurs.

Use the map below to answer question 18.

18. The map above shows three circles drawn around three different seismograph stations. The circles have radii equal to the distance between the seismograph station and the earthquake's epicenter. Which labeled point on the map represents the earthquake's epicenter? How do you know?
19. Compare and contrast the three types of seismic waves.
20. Draw three diagrams to show how each type of seismic wave moves through rocks.
21. Describe what happens to S-waves when they contact Earth's outer core. Decribe what happens to P-waves when they reach Earth's outer core. What is the *shadow zone*?

chapter 6

Volcanoes

chapter preview

sections
1. **Volcanoes and Earth's Moving Plates**
2. **Types of Volcanoes**
   Lab Identifying Types of Volcanoes
3. **Igneous Rock Features**
   Lab How do calderas form?
   Virtual Lab How does magma's composition affect a volcano's eruption?

Beautiful but Dangerous?

In late October, 2002, earthquakes triggered this vigorous eruption from Mt. Etna, a volcano on the Italian island of Sicily. Strombolian eruptions and lava fountains spewed gas, bombs, blocks, and liquid lava. Ash settled as far away as Libya.

Science Journal Do all volcanoes begin with violent, explosive eruptions? Write about your current beliefs, then do some research and write about your discoveries.

Start-Up Activities

Map a Volcano

You've seen pictures of volcanoes from the ground, but what would a volcano look like on a map? Volcanoes can be represented on maps that show the elevation of the land, topographic maps.

1. Obtain half of a foam ball from your teacher and place it on the top of a table with the flat side down.
2. Using a metric ruler and a permanent marker, mark 1-cm intervals on the foam ball. Start at the base of the ball and mark up at several places around the ball.
3. Connect the marks of equal elevation by drawing a line around the ball at the 1-cm mark, at the 2-cm mark, etc.
4. Look directly down on the top of the ball. Make a drawing of what you see in your Science Journal.
5. **Think Critically** In your Science Journal, write a paragraph that explains how your drawing shows the volcano's general shape.

Volcanoes Make the following Foldable to compare and contrast the characteristics of explosive and quiet volcanic eruptions.

STEP 1 Fold one sheet of paper lengthwise.

STEP 2 Fold into thirds.

STEP 3 Unfold and draw overlapping ovals. Cut the top sheet along the folds.

STEP 4 Label the ovals *Explosive Eruptions*, *Both*, and *Quiet Eruptions*, as shown.

Construct a Venn Diagram As you read the chapter, list the characteristics unique to explosive eruptions under the left tab, those unique to quiet eruptions under the right tab, and those characteristics common to both under the middle tab.

 Preview this chapter's content and activities at bookf.msscience.com

section 1

Volcanoes and Earth's Moving Plates

as you read

What You'll Learn
- **Describe** how volcanoes can affect people.
- **List** conditions that cause volcanoes to form.
- **Identify** the relationship between volcanoes and Earth's moving plates.

Why It's Important
Volcanoes can be dangerous to people and their communities.

Review Vocabulary
lava: molten rock material flowing from volcanoes onto Earth's surface

New Vocabulary
- volcano
- vent
- crater
- hot spot

What are volcanoes?

A **volcano** is an opening in Earth that erupts gases, ash, and lava. Volcanic mountains form when layers of lava, ash, and other material build up around these openings. Can you name any volcanoes? Did you know that Earth has more than 600 active volcanoes?

Most Active Volcanoes Kilauea (kee low AY ah), located in Hawaii, is the world's most active volcano. For centuries, this volcano has been erupting, but not explosively. In May of 1990, most of the town of Kalapana Gardens was destroyed, but no one was hurt because the lava moved slowly and people could escape. The most recent series of eruptions from Kilauea began in January 1983 and still continues.

The island country of Iceland is also famous for its active volcanoes. It sits on an area where Earth's plates move apart and is known as the land of fire and ice. The February 26, 2000, eruption of Hekla, in Iceland, is shown in **Figure 1**.

Figure 1 This photo of the February 26, 2000, eruption of Hekla shows why Iceland is known as the land of fire and ice.

Figure 2 This town on Montserrat was devastated by the eruption of Soufrière Hills volcano.

Effects of Eruptions

When volcanoes erupt, they often have direct, dramatic effects on the lives of people and their property. Lava flows destroy everything in their path. Falling volcanic ash can collapse buildings, block roads, and in some cases cause lung disease in people and animals. Sometimes, volcanic ash and debris rush down the side of the volcano. This is called a pyroclastic flow. The temperatures inside the flow can be high enough to ignite wood. When big eruptions occur, people often are forced to abandon their land and homes. People who live farther away from volcanoes are more likely to survive, but cities, towns, crops, and buildings in the area can be damaged by falling debris.

Human and Environmental Impacts The eruption of Soufrière (sew FREE er) Hills volcano in Montserrat, which began in July of 1995, was one of the largest recent volcanic eruptions near North America. Geologists knew it was about to erupt, and the people who lived near it were evacuated. On June 25, 1997, large pyroclastic flows swept down the volcano. As shown in **Figure 2,** they buried cities and towns that were in their path. The eruption killed 20 people who ignored the evacuation order.

When sulfurous gases from volcanoes mix with water vapor in the atmosphere, acid rain forms. The vegetation, lakes, and streams around Soufrière Hills volcano were impacted significantly by acid rain. As the vegetation died, shown in **Figure 3,** the organisms that lived in the forest were forced to leave or also died.

Figure 3 The vegetation near the volcano on Chances Peak, on the island of Montserrat in the West Indies, was destroyed by acid rain, heat, and ash.

Volcanologists
Volcanologists research many aspects of volcanoes, including space volcanoes. Io, a moon of Jupiter, has many active volcanoes. Become an amateur volcanologist; choose an aspect of volcanology that interests you, research your topic, and then create an exciting news broadcast to share the information with the class.

How do volcanoes form?

What happens inside Earth to create volcanoes? Why are some areas of Earth more likely to have volcanoes than others? Deep inside Earth, heat and pressure changes cause rock to melt, forming liquid rock or magma. Some deep rocks already are melted. Others are hot enough that a small rise in temperature or drop in pressure can cause them to melt and form magma. What makes magma come to the surface?

Magma Forced Upward Magma is less dense than the rock around it, so it is forced slowly toward Earth's surface. You can see this process if you turn a bottle of cold syrup upside down. Watch the dense syrup force the less dense air bubbles slowly toward the top.

Reading Check *Why is magma forced toward Earth's surface?*

After many thousands or even millions of years, magma reaches Earth's surface and flows out through an opening called a **vent.** As lava flows out, it cools quickly and becomes solid, forming layers of igneous rock around the vent. The steep-walled depression around a volcano's vent is the **crater. Figure 4** shows magma being forced out of a volcano.

Figure 4 This cutaway diagram shows how a volcano is formed and how magma from inside Earth is forced to the surface.

After magma reaches the surface, it is called lava. Ash, lava, and gases erupt from the crater of the volcano.

Pressure forces magma up the main pipe and any branch pipes that have formed.

Magma collects in a magma chamber underground before it is forced toward the surface.

Where do volcanoes occur?

Volcanoes often form in places where plates are moving apart, where plates are moving together, and at locations called hot spots. You can find locations of active volcanoes at plate boundaries and at hot spots on the map in **Figure 5.** Many examples can be found of volcanoes around the world that form at these three different kinds of areas. You'll explore volcanoes in Iceland, on the island of Montserrat, and in Hawaii.

Figure 5 This map shows the locations of volcanoes, hot spots, and plate boundaries around the world. The Ring of Fire is a belt of active volcanoes that circles the Pacific Ocean.

Divergent Plate Boundaries Iceland is a large island in the North Atlantic Ocean. It is near the Arctic Circle and therefore has some glaciers. Iceland has volcanic activity because it is part of the Mid-Atlantic Ridge.

Figure 6 This diagram shows how volcanic activity occurs where Earth's plates move apart.

The Mid-Atlantic Ridge is a divergent plate boundary, which is an area where Earth's plates are moving apart. When plates separate, they form long, deep cracks called rifts. Lava flows from these rifts and is cooled quickly by seawater. **Figure 6** shows how magma rises at rifts to form new volcanic rock. As more lava flows and hardens, it builds up on the seafloor. Sometimes, the volcanoes and rift eruptions rise above sea level, forming islands such as Iceland. In 1963, the new island Surtsey was formed during a volcanic eruption.

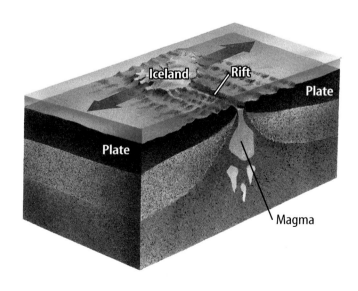

SECTION 1 Volcanoes and Earth's Moving Plates **F** ◆ **159**

Figure 7 Volcanoes can form where plates collide and one plate slides below the other.

Magma is less dense than rock, so it is forced upward and eventually erupts from the volcano.

As the oceanic plate slides downward, rock melts and forms magma.

Modeling Magma Movement

Procedure
1. Pour **water** into a **clear-plastic cup**.
2. Pour a small amount of **olive oil** into a separate plastic cup.
3. Extract a small amount of oil with a **dropper**.
4. Submerge the dropper tip into the water cup and slowly squeeze oil drops into the water.

Analysis
1. Describe what happened to the oil.
2. How do your observations compare with the movement of magma within Earth's crust?

Convergent Plate Boundaries Places where Earth's plates move together are called convergent plate boundaries. They include areas where an oceanic plate slides below a continental plate as in **Figure 7,** and where one oceanic plate slides below another oceanic plate. The Andes in South America began forming when an oceanic plate started sliding below a continental plate. Volcanoes that form on convergent plate boundaries tend to erupt more violently than other volcanoes do.

Magma forms when the plate sliding below another plate gets deep enough and hot enough to melt partially. The magma then is forced upward to the surface, forming volcanoes like Soufrière Hills on the island of Montserrat.

Hot Spots The Hawaiian Islands are forming as a result of volcanic activity. However, unlike Iceland, they haven't formed at a plate boundary. The Hawaiian Islands are in the middle of the Pacific Plate, far from its edges. What process could be forming them?

It is thought that some areas at the boundary between Earth's mantle and core are unusually hot. Hot rock at these areas is forced toward the crust where it melts partially to form a **hot spot.** The Hawaiian Islands sit on top of a hot spot under the Pacific Plate. Magma has broken through the crust to form several volcanoes. The volcanoes that rise above the water form the Hawaiian Islands, shown in **Figure 8.**

Figure 8 This satellite photo shows five of the Hawaiian Islands, which actually are volcanoes. **Explain** *why they are in a relatively straight line.*

This illustration shows that the Hawaiian Islands were formed over a hot spot.

The Hawaiian Islands As you can see in **Figure 8,** the Hawaiian Islands are all in a line. This is because the Pacific Plate is moving over a stationary hot spot. Kauai, the oldest Hawaiian island, was once located where the big island, Hawaii, is situated today. As the plate moved, Kauai moved away from the hot spot and became dormant. As the Pacific Plate continued to move, the islands of Oahu, Molokai, Maui, and Hawaii were formed. The Hawaiian Islands formed over a period of about 5 million years.

section 1 review

Summary

What are volcanoes?
- A volcano is an opening in Earth's surface that erupts gases, ash, and lava.

Effects of Eruptions
- Direct effects of volcanic eruptions can be caused by lava flows, pyroclastic flows, and falling ash.
- Volcanic eruptions also produce indirect effects, such as acid rain.

How do volcanoes form?
- Volcanoes form when magma is forced up and flows onto Earth's surface as lava.
- A crater is a steep-walled depression around a volcano's vent.

Where do volcanoes occur?
- Volcanoes form where one plate sinks beneath another plate, where two plates are moving apart, and at hot spots.

Self Check

1. **Explain** why volcanoes are commonly found at the edges of Earth's moving plates.
2. **Describe** what effects pyroclastic flows have on people.
3. **Explain** why lava cools rapidly along a mid-ocean ridge. How might underwater lava differ from surface lava?
4. **Describe** the processes that cause Soufrière Hills volcano to erupt.
5. **Think Critically** If the Pacific Plate stopped moving, what might happen to the Big Island of Hawaii?

Applying Skills

6. **Concept Map** Make a concept map that shows how the Hawaiian Islands formed. Use the following phrases: *volcano forms, plate moves, volcano becomes dormant,* and *new volcano forms.* Draw and label an illustration of this process.

bookf.msscience.com/self_check_quiz

section 2
Types of Volcanoes

as you read

What You'll Learn
- **Explain** how the explosiveness of a volcanic eruption is related to the silica and water vapor content of its magma.
- **List** three forms of volcanoes.

Why It's Important
If you know the type of volcano, you can predict how it will erupt.

Review Vocabulary
magma: hot, melted rock material beneath Earth's surface

New Vocabulary
- shield volcano
- tephra
- cinder cone volcano
- composite volcano

What controls eruptions?

Some volcanic eruptions are explosive, like those from Soufrière Hills volcano, Mount Pinatubo, and Mount St. Helens. In others, the lava quietly flows from a vent, as in the Kilauea eruptions. What causes these differences?

Two important factors control whether an eruption will be explosive or quiet. One factor is the amount of water vapor and other gases that are trapped in the magma. The second factor is how much silica is present in the magma. Silica is a compound composed of the elements silicon and oxygen.

Trapped Gases When you shake a soft-drink container and then quickly open it, the pressure from the gas in the drink is released suddenly, spraying the drink all over. In the same way, gases such as water vapor and carbon dioxide are trapped in magma by the pressure of the surrounding magma and rock. As magma nears the surface, it is under less pressure. This allows the gas to escape from the magma. Gas escapes easily from some magma during quiet eruptions. However, gas that builds up to high pressures eventually causes explosive eruptions such as the one shown in **Figure 9**.

Figure 9 Mount St. Helens erupted on May 18, 1980.

8:32 A.M.

38 seconds later

Water Vapor The magma at some convergent plate boundaries contains a lot of water vapor. This is because oceanic plate material and some of its water slide under other plate material at some convergent plate boundaries. The trapped water vapor in the magma can cause explosive eruptions.

Composition of Magma

The second major factor that affects the nature of the eruption is the composition of the magma. Magma can be divided into two major types—silica poor and silica rich.

Quiet Eruptions Magma that is relatively low in silica is called basaltic magma. It is fluid and produces quiet, nonexplosive eruptions such as those at Kilauea. This type of lava pours from volcanic vents and runs down the sides of a volcano. As this *pahoehoe* (pa-HOY-hoy) lava cools, it forms a ropelike structure. If the same lava flows at a lower temperature, a stiff, slowly moving *aa* (AH-ah) lava forms. In fact, you can walk right up to some aa lava flows on Kilauea.

Figure 10 shows some different types of lava. These quiet eruptions form volcanoes over hot spots such as the Hawaiian volcanoes. Basaltic magmas also flow from rift zones, which are long, deep cracks in Earth's surface. Many lava flows in Iceland are of this type. Because basaltic magma is fluid when it is forced upward in a vent, trapped gases can escape easily in a nonexplosive manner, sometimes forming lava fountains. Lavas that flow underwater form pillow lava formations. They consist of rock structures shaped like tubes, balloons, or pillows.

Topic: Kilauea Volcano
Visit bookf.msscience.com for Web links to information about Kilauea volcano in Hawaii.

Activity On a map of the Hawaiian Islands, identify the oldest and most recent islands. Next, indicate where Kilauea volcano is located on the Big Island. Do you see a directional pattern? Indicate on your Hawaiian map where you believe the next Hawaiian island will form.

42 seconds later

53 seconds later

NATIONAL GEOGRAPHIC VISUALIZING LAVA

Figure 10

Lava rarely travels faster than a few kilometers an hour. Therefore, it poses little danger to people. However, homes and property can be damaged. On land, there are two main types of lava flows—aa and pahoehoe. When lava comes out of cracks in the ocean floor, it is called pillow lava. The lava cooling here came from a volcanic eruption on the island of Hawaii.

Aa flows, like this one on Mount Etna in Italy, carry sharp angular chunks of rock called scoria. Aa flows move slowly and are intensely hot.

Pillow lava occurs where lava oozes out of cracks in the ocean floor. It forms pillow-shaped lumps as it cools. Pillow lava is the most common type of lava on Earth.

Pahoehoe flows, like this one near Kilauea's Mauna Ulu Crater in Hawaii, are more fluid than aa flows. They develop a smooth skin and form ropelike patterns when they cool.

Figure 11 Magmas that are rich in silica produce violent eruptions.

Violent eruptions, such as this one in Alaska, often produce a lot of volcanic ash.

This color enhanced view of volcanic ash, from a 10 million year old volcano in Nebraska, shows the glass particles that make up ash.

Explosive Magma Silica-rich, or granitic, magma produces explosive eruptions such as those at Soufrière Hills volcano. This magma sometimes forms where Earth's plates are moving together and one plate slides under another. As the plate that is sliding under the other goes deeper, some rock is melted. The magma is forced upward by denser surrounding rock, comes in contact with the crust, and becomes enriched in silica. Silica-rich granitic magma is thick, and gas gets trapped inside, causing pressure to build up. When an explosive eruption occurs, as shown in **Figure 11,** the gases expand rapidly, often carrying pieces of lava in the explosion.

Reading Check *What type of magmas produce violent eruptions?*

Some magmas have an andesitic composition. Andesitic magma is more silica rich than basaltic magma is, but it is less silica rich than granitic magma. It often forms at convergent plate boundaries where one plate slides under the other. Because of their higher silica content, they also erupt more violently than basaltic magmas. One of the biggest eruptions in recorded history, Krakatau, was primarily andesitic in composition. The word *andesitic* comes from the Andes, which are mountains located along the western edge of South America, where andesite rock is common. Many of the volcanoes encircling the Pacific Ocean also are made of andesite.

Volcanic Ash When volcanoes erupt, ash often is spread over a great distance. People who live near volcanoes must be careful not to inhale too much of the ash particles because the particles can cause respiratory problems. In your Science Journal, describe what people can do to prevent exposure to volcanic ash.

Mini LAB

Modeling Volcanic Cones

Procedure
1. Pour **dry sand** or **sugar** onto one spot on a **paper plate**. **WARNING:** *Do not taste, eat, or drink any materials used in the lab.*
2. Mix a batch of **plaster of paris** and pour it onto one spot on another paper plate.
3. Allow the plaster of paris to dry. Use a **protractor** to measure the slope angles of the sides of the volcanoes.

Analysis
What form of volcano is represented by the model with steeper sides?

Forms of Volcanoes

A volcano's form depends on whether it is the result of a quiet or an explosive eruption and the type of lava it is made of—basaltic, granitic, or andesitic (intermediate). The three basic types of volcanoes are shield volcanoes, cinder cone volcanoes, and composite volcanoes.

Shield Volcano Quiet eruptions of basaltic lava spread out in flat layers. The buildup of these layers forms a broad volcano with gently sloping sides called a **shield volcano,** as seen in **Figure 12.** The Hawaiian Islands are examples of shield volcanoes. Basaltic lava also can flow onto Earth's surface through large cracks called fissures. This type of eruption forms flood basalts, not volcanoes, and accounts for the greatest volume of erupted volcanic material. The basaltic lava flows over Earth's surface, covering large areas with thick deposits of basaltic igneous rock when it cools. The Columbia Plateau located in the northwestern United States was formed in this way. Much of the new seafloor that originates at mid-ocean ridges forms as underwater flood basalts.

Cinder Cone Volcano Explosive eruptions throw lava and rock high into the air. Bits of rock or solidified lava dropped from the air are called **tephra** (TEH fruh). Tephra varies in size from volcanic ash, to cinders, to larger rocks called bombs and blocks. When tephra falls to the ground, it forms a steep-sided, loosely packed **cinder cone volcano,** as seen in **Figure 13.**

Figure 12 A shield volcano like Mauna Loa, shown here, is formed when lava flows from one or more vents without erupting violently.

Paricutín On February 20, 1943, a Mexican farmer learned about cinder cones when he went to his cornfield. He noticed that a hole in his cornfield that had been there for as long as he could remember was giving off smoke. Throughout the night, hot glowing cinders were thrown high into the air. In just a few days, a cinder cone several hundred meters high covered his cornfield. This is the volcano named Paricutín.

Composite Volcano Some volcanic eruptions can vary between quiet and violent, depending on the amount of trapped gases and how rich in silica the magma is. An explosive period can release gas and ash, forming a tephra layer. Then, the eruption can switch to a quieter period, erupting lava over the top of the tephra layer. When this cycle of lava and tephra is repeated over and over in alternating layers, a **composite volcano** is formed. Composite volcanoes, shown in **Figure 14,** are found mostly where Earth's plates come together and one plate slides below the other. Soufrière Hills volcano is an example. As you can see in **Table 1** on the next page, many things affect eruptions and the form of a volcano.

Figure 13 Paricutín is a large, cinder cone volcano located in Mexico.

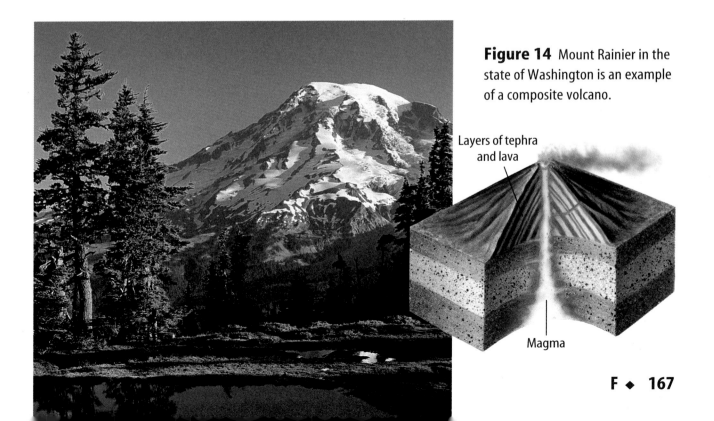

Figure 14 Mount Rainier in the state of Washington is an example of a composite volcano.

Violent Eruptions Soufrière Hills volcano formed as ocean floor of the North American Plate and the South American Plate slid beneath the Caribbean Plate, causing magma to form. Successive eruptions of lava and tephra produced the majestic composite volcanoes that tower above the surrounding landscape on Montserrat and other islands in the Lesser Antilles. Before the 1995 eruption, silica-rich magma rose and was trapped beneath the surface. As the magma was forced toward Earth's surface, the pressure on the underlying magma was released. This started a series of eruptions that were still continuing in the year 2003.

Table 1 Thirteen Selected Eruptions

Volcano and Location	Year	Type	Eruptive Force	Magma Content		Ability of Magma to Flow	Products of Eruption
				Silica	H$_2$O		
Mount Etna, Sicily	1669	composite	moderate	high	low	medium	lava, ash
Tambora, Indonesia	1815	cinder cone	high	high	high	low	cinders, ash
Krakatau, Indonesia	1883	composite	high	high	high	low	cinders, ash
Mount Pelée, Martinique	1902	cinder cone	high	high	high	low	gas, ash
Vesuvius, Italy	1906	composite	moderate	high	low	medium	lava, ash
Mount Katmai, Alaska	1912	composite	high	high	high	low	lava, ash, gas
Paricutín, Mexico	1943	cinder cone	moderate	high	low	medium	ash, cinders
Surtsey, Iceland	1963	shield	moderate	low	low	high	lava, ash
Mount St. Helens, Washington	1980	composite	high	high	high	low	gas, ash
Kilauea, Hawaii	1983	shield	low	low	low	high	lava
Mount Pinatubo, Philippines	1991	composite	high	high	high	low	gas, ash
Soufrière Hills, Montserrat	1995	composite	high	high	high	low	gas, ash, rocks
Popocatépetl, Mexico	2000	composite	moderate	high	low	medium	gas, ash

Figure 15 Not much was left after Krakatau erupted in 1883.

Krakatau One of the most violent eruptions in recent times occurred on an island in the Sunda Straits near Indonesia in August of 1883. Krakatau, a volcano on the island, erupted with such force that the island disappeared, as shown in **Figure 15A**. Most of the island collapsed into the emptied magma chamber. The noise of the eruption was so loud that it woke people in Australia and was heard as far away as 4,653 km from the island. Ash from the eruption fell in Singapore, which is 840 km to the north, and the area around the volcano was in complete darkness for 24 h. More than 36,000 people were killed, most by the giant tsunami waves created by the eruption. Global temperatures were lowered as much as 1.2°C by particles blown into the atmosphere and didn't return to normal until 1888.

section 2 review

Summary

What controls eruptions?
- The amount of water vapor and other gases control the type of eruption and the amount of silica present in the magma.

Composition of Magma
- Magma can be divided into two major types—silica rich and silica poor.

Forms of Volcanoes
- A shield volcano is a broad, gently sloping volcano formed by quiet eruptions of basaltic lava.
- A cinder cone volcano is a steep-sided, loosely packed volcano formed from tephra.
- Composite volcanoes are formed by alternating explosive and quiet eruptions that produce layers of tephra and lava.

Self Check

1. **Define** the term *tephra,* and where it can be found.
2. **Describe** the differences between basaltic and granitic magma.
3. **Identify** the specific water vapor and silica conditions that cause differences in eruptions.
4. **Describe** how the Hawaiian Islands formed.
5. **Think Critically** In 1883, Krakatau in Indonesia erupted. Infer which kind of lava Krakatau erupted—lava rich in silica or lava low in silica. Support your inference using data in **Table 1**.

Applying Skills

6. **Compare and contrast** Kilauea and Mount Pinatubo using information from **Table 1**.

Identifying Types of Volcanoes

You have learned that certain properties of magma are related to the type of eruption and the form of the volcano that will develop. Do this lab to see how to make and use a table that relates the properties of magma to the form of volcano that develops.

◗ Real-World Question

Are the silica and water content of a magma related to the form of volcano that develops?

Goals
- **Determine** any relationship between the ability of magma to flow and eruptive force.
- **Determine** any relationship between magma composition and eruptive force.

Materials
Table 1 (thirteen selected eruptions)
paper
pencil

◗ Procedure

1. Copy the graph shown above.
2. Using the information from **Table 1,** plot the magma content for each of the volcanoes listed by writing the name of the basic type of volcano in the correct spot on the graph.

3. What relationship appears to exist between the silica or water content of the magma and the nature of the material ejected from the volcano?
4. How is the ability of a magma to flow related to its silica content?
5. **Infer** which of the two variables, silica or water content, appears to have the greater effect on the eruptive force of the volcano.
6. **Describe** the relationship that appears to exist between the silica and water content of the magma and the type of volcano that is produced.

◗ Conclude and Apply

1. What relationship appears to exist between the ability of the magma to flow and the eruptive force of the volcano?
2. Which would be more liquidlike: magma that flows easily or magma that flows with difficulty?

Communicating Your Data

Create a flowchart that shows the relationship between magma composition and the type of volcano formed. **For more help, refer to the** Science Skill Handbook.

section 3
Igneous Rock Features

Intrusive Features

You can observe volcanic eruptions because they occur at Earth's surface. However, far more activity occurs underground. In fact, most magma never reaches Earth's surface to form volcanoes or to flow as flood basalts. This magma cools slowly underground and produces underground rock bodies that could become exposed later at Earth's surface by erosion. These rock bodies are called intrusive igneous rock features. There are several different types of intrusive features. Some of the most common are batholiths, sills, dikes, and volcanic necks. What do intrusive igneous rock bodies look like? You can see illustrations of these features in **Figure 16.**

as you read

What **You'll Learn**
- **Describe** intrusive igneous rock features and how they form.
- **Explain** how a volcanic neck and a caldera form.

Why **It's Important**
Many features formed underground by igneous activity are exposed at Earth's surface by erosion.

Review Vocabulary
intrude: to enter by force; cut in
extrude: to force or push out

New Vocabulary
- batholith
- dike
- sill
- volcanic neck
- caldera

Figure 16 This diagram shows intrusive and other features associated with volcanic activity.
Identify *which features shown are formed above ground. Which are formed by intrusive activities?*

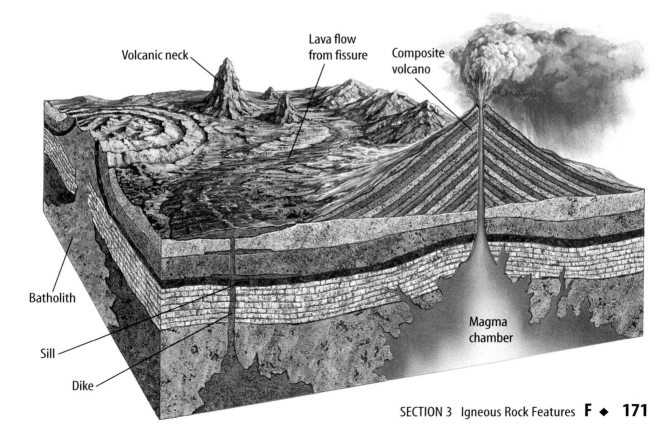

SECTION 3 Igneous Rock Features **F** ◆ **171**

Batholiths The largest intrusive igneous rock bodies are **batholiths.** They can be many hundreds of kilometers in width and length and several kilometers thick. Batholiths form when magma bodies that are being forced upward from inside Earth cool slowly and solidify before reaching the surface. However, not all of them remain hidden inside Earth. Some batholiths have been exposed at Earth's surface by many millions of years of erosion. The granite domes of Yosemite National Park are the remains of a huge batholith that stretches across much of the length of California.

Applying Math — Calculate Percent

CLASSIFYING IGNEOUS ROCKS Igneous rocks are classified into three types depending on the amount of silica they contain. Basaltic rocks contain approximately 45 percent to 52 percent silica. Andesitic, or intermediate, rocks contain about 52 percent to 66 percent silica, and granitic rocks have more than 66 percent silica. The lighter the color is, the higher the silica content is. A 900-kg block of igneous rock contains 630 kg of silica. Calculate the percent of silica in the rock to classify it.

Solution

1 This is what you know:
- rock = 900 kg
- silica = 630 kg

2 This is what you need to find: The percentage of silica: x

3 This is the equation you need to use: Mass of silica / mass of rock = x / 100

4 Solve the equation for x:
- $x = (630 \text{ kg}/900 \text{ kg}) \times 100$
- $x = 70$ percent, therefore, the rock is granitic.

Check your answer by dividing it by 100, then multiplying by 900. Did you get the given amount of silica?

Practice Problems

1. A 250-kg boulder of basalt contains 125 kg of silica. Use the classification system to determine whether basalt is light or dark.

2. Andesite is an intermediate, medium-colored rock with a silica content ranging from 52 percent to 66 percent. About how many kilograms of silica would you predict to be in a 68-kg boulder of andesite?

 For more practice, visit bookf.msscience.com/math_practice

Dikes and Sills Magma sometimes squeezes into cracks in rock below the surface. This is like squeezing toothpaste into the spaces between your teeth. Magma that is forced into a crack that cuts across rock layers and hardens is called a **dike.** Magma that is forced into a crack parallel to rock layers and hardens is called a **sill.** These features are shown in **Figure 17.** Most dikes and sills run from a few meters to hundreds of meters long.

Other Features

When a volcano stops erupting, the magma hardens inside the vent. Erosion, usually by water and wind, begins to wear away the volcano. The cone is much softer than the solid igneous rock in the vent. Thus, the cone erodes first, leaving behind the solid igneous core as a **volcanic neck.** Ship Rock in New Mexico, shown in **Figure 17,** is a good example of a volcanic neck.

Science Online

Topic: Igneous Rock Features
Visit bookf.msscience.com for Web links to information about igneous rock features.

Activity Create a collage for artistic competition by using a variety of pictures of igneous rock features. For extra challenge, research Devils Tower, Wyoming. Develop your own hypothesis for its formation, and present your ideas as a panel discussion with other classmates.

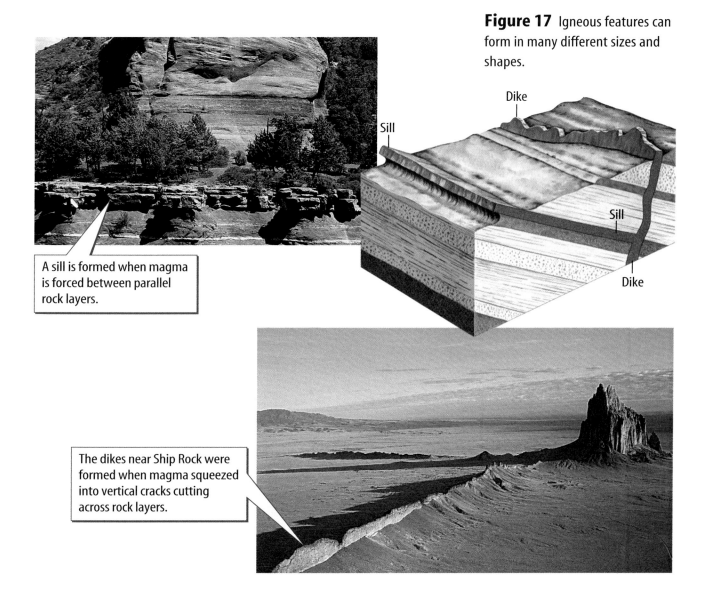

Figure 17 Igneous features can form in many different sizes and shapes.

A sill is formed when magma is forced between parallel rock layers.

The dikes near Ship Rock were formed when magma squeezed into vertical cracks cutting across rock layers.

Figure 18 Calderas form when the top of a volcano collapses.

Calderas Sometimes after an eruption, the top of a volcano can collapse, as seen in **Figure 18.** This produces a large depression called a **caldera.** Crater Lake in Oregon, shown in **Figure 19,** is a caldera that filled with water and is now a lake. Crater Lake formed after the violent eruption and destruction of Mount Mazama about 7,000 years ago.

Magma is forced upward, causing volcanic activity to occur.

The magma chamber partially empties, causing rock to collapse into the emptied chamber below the surface. This forms a circular-shaped caldera.

Crater Lake in Oregon formed when water collected in the circular space left when surface material collapsed.

Figure 19 Wizard Island in Crater Lake is a cinder cone volcano that erupted after the formation of the caldera.
Explain *what causes a caldera to form.*

Igneous Features Exposed You have learned in this chapter that Earth's surface is built up and worn down continually. The surface of Earth is built up by volcanoes. Also, igneous rock is formed when magma hardens below ground. Eventually, the processes of weathering and erosion wear down rock at the surface, exposing features like batholiths, dikes, and sills.

section 3 review

Summary

Intrusive Features

- Intrusive igneous rock features are formed from magma that is forced upward toward Earth's crust, then slowly cools and solidifies underground before reaching the surface.
- Batholiths, dikes, and sills are a few examples of intrusive igneous rock features.

Other Features

- A volcanic neck is the solid igneous core of a volcano left behind after the softer cone has been eroded.
- A caldera is a large, circular-shaped depression that forms when the top of a volcano collapses.

Self Check

1. **Compare and contrast** a caldera and a crater.
2. **Illustrate** how a sill forms. How is it different from a dike?
3. **Describe** a batholith and explain how it forms.
4. **Think Critically** Why are the large, granite dome features of Yosemite National Park in California considered to be intrusive volcanic features when they are exposed at the surface?

Applying Math

5. **Calculate** Basaltic rocks contain approximately 45 percent to 52 percent silica. About how many kilograms of silica would you predict to be in a 68-kg boulder of basalt?

LAB Design Your Own

How do calderas form?

Goals
- **Design** a volcano setup that will demonstrate how a caldera could form.
- **Observe** what happens during trials with your volcano setup.
- **Describe** what you observe.

Possible Materials
small box
small balloon
paper
newspaper
flour
plastic tubing
clamp for tubing
tape
scissors

Safety Precautions

Real-World Question
A caldera is a depression that forms when the top of a volcano collapses after an eruption. What might cause the top of a volcano to collapse?

Form a Hypothesis
Based on your reading about volcanoes, state a hypothesis about what would happen if the magma inside the magma chamber of a volcano were suddenly removed.

Test Your Hypothesis
Make a Plan
1. As a group, agree upon the hypothesis and identify which results will support the hypothesis.
2. **Design** a volcano that allows you to test your hypothesis. What materials will you use to build your volcano?
3. What will you remove from inside your volcano to represent the loss of magma? How will you remove it?
4. Where will you place your volcano? What will you do to minimize messes?
5. **Identify** all constants, variables, and controls of the experiment.

Follow Your Plan
1. Make sure your teacher approves your plan before you start.
2. **Construct** your volcano with any features that will be required to test your hypothesis.
3. **Conduct** one or more appropriate trials to test your hypothesis. Record any observations that you make and any other data that are appropriate to test your hypothesis.

Using Scientific Methods

◉ Analyze Your Data

1. **Describe** in words or with a drawing what your volcano looked like before you began.
2. **Observe** what happened to your volcano during the experiment that you conducted? Did its appearance change?
3. **Describe** in words or with a drawing what your volcano looked like after the trial.
4. **Observe** What other observations did you make?
5. **Describe** any other data that you recorded.

◉ Conclude and Apply

1. **Draw Conclusions** Did your observations support your hypothesis? Explain.
2. **Explain** how your demonstration was similar to what might happen to a real volcano. How was it different?

Communicating Your Data

Make a 4-sequence time-lapse diagram with labels and descriptions of how a caldera forms. Use your visual aid to describe caldera formation to students in another class.

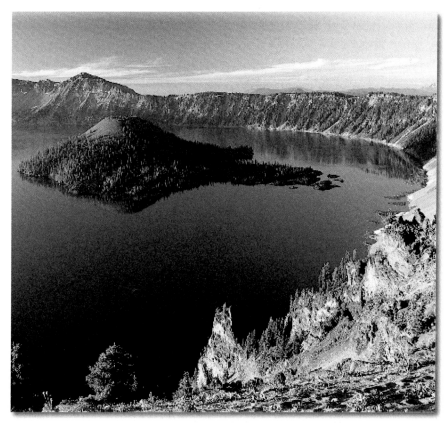

LAB F ◆ 177

Oops! Accidents in SCIENCE

SOMETIMES GREAT DISCOVERIES HAPPEN BY ACCIDENT!

Buried in Ash

A long-forgotten city is accidentally found after 2,000 years

In the heat of the Italian Sun, a farmer digs a new well for water. He thrusts his shovel into the ground one more time. But instead of hitting water, the shovel strikes something hard; a slab of smooth white marble.

Under the ground lay the ancient city of Herculaneum (her kew LAY nee um). The city, and its neighbor Pompeii (pom PAY) had been buried for more than 1,600 years. On August 24, 79 A.D., Mount Vesuvius, a nearby volcano, erupted and buried both cities with pumice, rocks, mud, and ash.

Back in Time

The Sun shone over the peaceful town of Herculaneum on that August morning almost 2,000 years ago. But at about 1 P.M., that peace was shattered forever.

With massive force, the peak of Vesuvius exploded, sending six cubic kilometers of ash and pumice into the sky. Hours later, a fiery surge made its way from the volcano to the city. These pyroclastic flows continued as more buildings were crushed and buried by falling ash and pumice. Within six hours, much of the city was totally buried under the flows. After six surges from Vesuvius, the deadly eruption ceased. But the city had disappeared under approximately 21 m of ash, rock, and mud.

A City Vanishes

More than 3,600 people were killed in the natural disaster. Scientists believe that many died trying to protect their faces from the pyroclastic surges that filled the air with hot ash. Those able to escape returned to find no trace of their city. Over hundreds of years, grass and fields covered Herculaneum, erasing it from human memory.

Archaeologists have unearthed perfectly preserved mosaics and a library with ancient scrolls in excellent condition. Archaeologists found skeletons and voids that were filled with plaster to form casts of people who died the day Vesuvius erupted. Visitors to the site can see a Roman woman, a teen-aged girl, and a soldier with his sword still in his hand.

Much of Herculaneum still lies buried beneath thick layers of volcanic ash, and archaeologists still are digging to expose more of the ruins. Their work is helping scientists better understand everyday life in an ancient Italian town. But, if it weren't for a farmer's search for water, Herculaneum might not have been discovered at all!

Excavated ruins with Mount Vesuvius in the background

Research the history of your town. Ask your local librarian to help "unearth" maps, drawings, or photos that let you travel back in time! Design a two-layer map that shows the past and the present.

Science online
For more information, visit bookf.msscience.com/oops

chapter 6 Study Guide

Reviewing Main Ideas

Section 1: Volcanoes and Earth's Moving Plates

1. Volcanoes can be dangerous to people because they can cause deaths and destroy property.
2. Rocks in the crust and mantle melt to form magma, which is forced toward Earth's surface. When the magma flows through vents, it's called lava and forms volcanoes.
3. Volcanoes can form over hot spots or when Earth's plates pull apart or come together.

Section 2: Types of Volcanoes

1. The three types of volcanoes are composite volcanoes, cinder cone volcanoes, and shield volcanoes.
2. Shield volcanoes produce quiet eruptions. Cinder cone and composite volcanoes can produce explosive eruptions.
3. Some lavas are thin and flow easily, producing quiet eruptions. Other lavas are thick and stiff, producing violent eruptions.

Section 3: Igneous Rock Features

1. Intrusive igneous rock bodies such as batholiths, dikes, and sills form when magma solidifies underground.
2. Batholiths are the most massive igneous rock bodies. Dikes and sills form when magma squeezes into cracks.
3. A caldera forms when the top of a volcano collapses, forming a large depression.

Visualizing Main Ideas

Copy and complete the following concept map on types of volcanic eruptions.

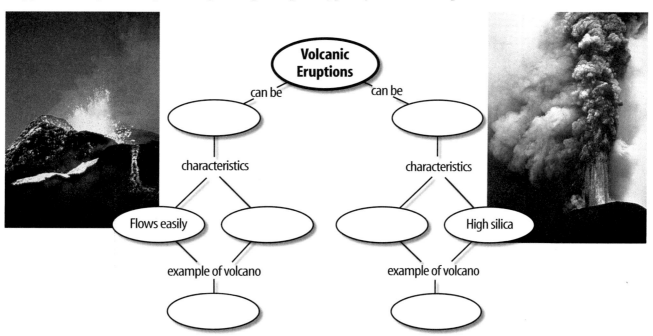

chapter 6 Review

Using Vocabulary

batholith p. 162
caldera p. 174
cinder cone volcano p. 166
composite volcano p. 167
crater p. 158
dike p. 173
hot spot p. 160
shield volcano p. 166
sill p. 173
tephra p. 166
vent p. 158
volcanic neck p. 173
volcano p. 156

Fill in the blanks with the correct vocabulary word or words.

1. A broad volcano with gently sloping sides is called a(n) _____.

2. Bits of rock or solidified lava dropped from the air after a volcanic eruption are _____.

3. Magma squeezed into a horizontal crack between rock layers is a(n) _____.

4. The steep-walled depression around a volcano's vent is called a(n) _____.

5. Magma squeezed into a vertical crack across rock layers is called a(n) _____.

Checking Concepts

Choose the word or phrase that best answers the question.

6. What type of boundary is associated with composite volcanoes?
   A) plates moving apart
   B) plates sticking and slipping
   C) plates moving together
   D) plates sliding past each other

7. Why is Hawaii made of volcanoes?
   A) Plates are moving apart.
   B) A hot spot exists.
   C) Plates are moving together.
   D) Rift zones exist.

8. What kind of magmas produce violent volcanic eruptions?
   A) those rich in silica
   B) those that are fluid
   C) those forming shield volcanoes
   D) those rich in iron

9. Magma that is low in silica generally produces what kind of eruptions?
   A) thick   C) quiet
   B) caldera   D) explosive

Use the photo below to answer question 10.

10. Which type of volcano, shown above, is made entirely of tephra?
    A) shield   C) cinder cone
    B) caldera   D) composite

11. What kind of volcano is Kilauea?
    A) shield   C) cinder cone
    B) composite   D) caldera cone

12. What is the largest intrusive igneous rock body?
    A) dike   C) sill
    B) volcanic neck   D) batholith

13. What is the process that formed Soufrière Hills volcano on Montserrat?
    A) plates sticking and slipping
    B) caldera formation
    C) plates sliding sideways
    D) plates moving together

chapter 6 Review

Thinking Critically

14. **Explain** how glaciers and volcanoes can exist on Iceland.

15. **Describe** what kind of eruption is produced when basaltic lava that is low in silica flows from a volcano.

16. **Explain** how volcanoes are related to earthquakes.

17. **Infer** Misti is a volcano in Peru. Peru is on the western edge of South America. How might this volcano have formed?

18. **Describe** the layers of a composite volcano. Which layers represent violent eruptions?

19. **Classify** the volcano Fuji, which has steep sides and is made of layers of silica-rich lava and ash.

Use the map below to answer question 20.

20. **Interpret Scientific Illustrations** Look at the map above. The Hawaiian Islands and Emperor Seamounts were formed when the Pacific Plate moved over a fixed hot spot. If the Emperor chain trends in a direction different from the Hawaiian Islands, what can you infer about the Pacific Plate?

21. **Concept Map** Make a network-tree concept map about where volcanoes can occur. Include the following words and phrases: *hot spots, divergent plate boundaries, convergent plate boundaries, volcanoes, can occur, examples, Iceland, Soufrière Hills,* and *Hawaiian Islands.*

Performance Activities

22. **Poster** Make a Venn diagram of the three basic types of volcanoes. Label them and indicate what cone formation, lava composition, eruption, and geologic location are expected of each type of volcano.

Applying Math

23. **Sea Level** The base of the volcano Mauna Loa is about 5,000 m below sea level. The total height of the volcano is 9,170 m. What percentage of the volcano is above sea level? Below sea level?

Use the table below to answer questions 24 and 25.

Volcano	Year of Eruption	Amount of Material Ejected
Tambora	1815	131 km^3
Katmai	1912	30 km^3
Novarupta	1912	15 km^3
Mt. St. Helens	1980	1.3 km^3
Pinatubo	1991	5.5 km^3

24. **Ejected Material** How many times greater was the volume of ejected material from Tambora, as compared to Mt. St. Helens?

25. **Graph** Design a bar graph to show the amount of ejected material from the volcanoes. Present the information from least to greatest volume.

Chapter 6 Standardized Test Practice

Part 1 Multiple Choice

Record your answers on the answer sheet provided by your teacher or on a sheet of paper.

Use the photo below to answer question 1.

1. Which of the following terms best describes the rock in the photo above?
   A. aa
   B. pahoehoe
   C. pillow lava
   D. ash

2. Which of the following is made of layers of ash and cooled lava flows?
   A. shield volcano
   B. plateau basalts
   C. composite volcano
   D. cinder cone volcano

3. Which of the following volcanoes is located in the United States?
   A. Hekla
   B. Paricutin
   C. Mount Vesuvius
   D. Mount St. Helens

4. Which of the following igneous features is parallel to the rock layers that it intrudes?
   A. batholith
   B. volcanic neck
   C. sill
   D. dike

5. Which of the following forms when the top of a volcano collapses into a partially emptied magma chamber?
   A. fissure
   B. crater
   C. caldera
   D. volcanic neck

Test-Taking Tip

Relax Stay calm during the test. If you feel yourself getting nervous, close your eyes and take five slow, deep breaths.

Use the graph below to answer questions 6 and 7.

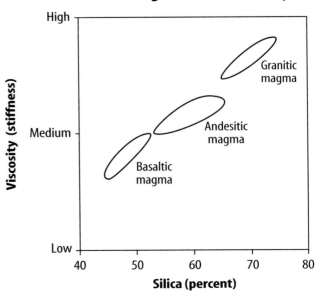

6. What relationship can be inferred from the graph?
   A. Magmas that have more silica are more viscous.
   B. Magmas that have less silica are more viscous.
   C. Magmas always have low viscosity.
   D. There is no relationship between silica content and viscosity.

7. What is the percentage of silica in Granitic magma?
   A. less than 45%
   B. 45–52%
   C. 53–65%
   D. greater than 65%

8. Which of the following is the finest type of tephra?
   A. volcanic ash
   B. volcanic bombs
   C. volcanic cinders
   D. volcanic blocks

Standardized Test Practice

Part 2 — Short Response/Grid In

Record your answers on the answer sheet provided by your teacher or on a sheet of paper.

9. What is a hot spot? Why do volcanoes often form at hot spots?

10. Why are the Hawaiian Islands in a line?

11. How is a dike different from a sill? Support your answer with a Venn diagram.

Use the table below to answer questions 12–14.

Eruption	Volume Percent Water Vapor
1	58.7
2	60.1
3	61.4
4	59.3
5	59.6

12. Calculate the mean, median, and range of the water vapor data in the table? Describe how this information would be helpful to a volcanologist.

13. Using the mean value that you calculated in question 12, what percentage of the volcanic gas consists of gases other than water vapor?

14. The water vapor content of Kilauea is above average when compared to other volcanoes. How might these data help to explain why lava fountains often occur on Kilauea?

15. What is the difference between magma and lava?

16. Explain how igneous rock forms from lava.

Part 3 — Open Ended

Record your answers on a sheet of paper.

17. Explain how volcanic necks, such as Ship Rock, form. Support your answer with a labeled diagram.

18. How does tephra form?

19. Why do some volcanoes occur where one plate sinks beneath another plate? Support your answer with a labeled diagram.

20. How can pillow-shaped bodies form from lava?

Use the map below to answer questions 21 and 22.

21. What kind of magma was required to create the Columbia Plateau as compared to Mt. St. Helens, only 200 miles away? What would you predict would be the percentage of water vapor in the two types of magma?

22. Where do volcanoes occur in the United States?

23. How do volcanoes affect people and their property? List four safety precautions for people living in volcanic areas.

Student Resources

CONTENTS

Science Skill Handbook 186

Scientific Methods 186
- Identify a Question 186
- Gather and Organize Information 186
- Form a Hypothesis 189
- Test the Hypothesis 190
- Collect Data 190
- Analyze the Data 193
- Draw Conclusions 194
- Communicate 194

Safety Symbols 195

Safety in the Science Laboratory 196
- General Safety Rules 196
- Prevent Accidents 196
- Laboratory Work 196
- Laboratory Cleanup 197
- Emergencies 197

Extra Try at Home Labs 198
- Panning Minerals 198
- Changing Rocks 198
- Why recycle? 199
- Measuring Movement 199
- Making Waves 200
- Mini Eruptions 200

Technology Skill Handbook ...201

Computer Skills 201
- Use a Word Processing Program ...201
- Use a Database 202
- Use the Internet 202
- Use a Spreadsheet 203
- Use Graphics Software 203

Presentation Skills 204
- Develop Multimedia Presentations 204
- Computer Presentations 204

Math Skill Handbook 205

Math Review 205
- Use Fractions 205
- Use Ratios 208
- Use Decimals 209
- Use Proportions 209
- Use Percentages 210
- Solve One-Step Equations 210
- Use Statistics 211
- Use Geometry 212

Science Applications 215
- Measure in SI 215
- Dimensional Analysis 215
- Precision and Significant Digits ...217
- Scientific Notation 217
- Make and Use Graphs 218

Reference Handbooks 220

Weather Map Symbols 220
Topographic Map Symbols 221
Periodic Table of the Elements 222
Minerals 224
Rocks 226

English/Spanish Glossary 227

Index 234

Credits 240

Science Skill Handbook

Scientific Methods

Scientists use an orderly approach called the scientific method to solve problems. This includes organizing and recording data so others can understand them. Scientists use many variations in this method when they solve problems.

Identify a Question

The first step in a scientific investigation or experiment is to identify a question to be answered or a problem to be solved. For example, you might ask which gasoline is the most efficient.

Gather and Organize Information

After you have identified your question, begin gathering and organizing information. There are many ways to gather information, such as researching in a library, interviewing those knowledgeable about the subject, testing and working in the laboratory and field. Fieldwork is investigations and observations done outside of a laboratory.

Researching Information Before moving in a new direction, it is important to gather the information that already is known about the subject. Start by asking yourself questions to determine exactly what you need to know. Then you will look for the information in various reference sources, like the student is doing in **Figure 1.** Some sources may include textbooks, encyclopedias, government documents, professional journals, science magazines, and the Internet. Always list the sources of your information.

Figure 1 The Internet can be a valuable research tool.

Evaluate Sources of Information Not all sources of information are reliable. You should evaluate all of your sources of information, and use only those you know to be dependable. For example, if you are researching ways to make homes more energy efficient, a site written by the U.S. Department of Energy would be more reliable than a site written by a company that is trying to sell a new type of weatherproofing material. Also, remember that research always is changing. Consult the most current resources available to you. For example, a 1985 resource about saving energy would not reflect the most recent findings.

Sometimes scientists use data that they did not collect themselves, or conclusions drawn by other researchers. This data must be evaluated carefully. Ask questions about how the data were obtained, if the investigation was carried out properly, and if it has been duplicated exactly with the same results. Would you reach the same conclusion from the data? Only when you have confidence in the data can you believe it is true and feel comfortable using it.

Science Skill Handbook

Interpret Scientific Illustrations As you research a topic in science, you will see drawings, diagrams, and photographs to help you understand what you read. Some illustrations are included to help you understand an idea that you can't see easily by yourself, like the tiny particles in an atom in **Figure 2.** A drawing helps many people to remember details more easily and provides examples that clarify difficult concepts or give additional information about the topic you are studying. Most illustrations have labels or a caption to identify or to provide more information.

Network Tree A type of concept map that not only shows a relationship, but how the concepts are related is a network tree, shown in **Figure 3.** In a network tree, the words are written in the ovals, while the description of the type of relationship is written across the connecting lines.

When constructing a network tree, write down the topic and all major topics on separate pieces of paper or notecards. Then arrange them in order from general to specific. Branch the related concepts from the major concept and describe the relationship on the connecting line. Continue to more specific concepts until finished.

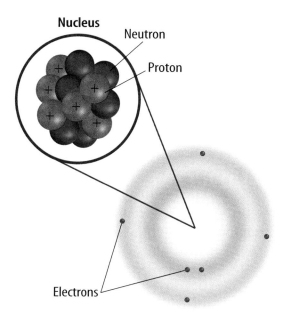

Figure 2 This drawing shows an atom of carbon with its six protons, six neutrons, and six electrons.

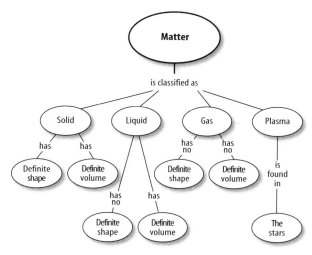

Figure 3 A network tree shows how concepts or objects are related.

Concept Maps One way to organize data is to draw a diagram that shows relationships among ideas (or concepts). A concept map can help make the meanings of ideas and terms more clear, and help you understand and remember what you are studying. Concept maps are useful for breaking large concepts down into smaller parts, making learning easier.

Events Chain Another type of concept map is an events chain. Sometimes called a flow chart, it models the order or sequence of items. An events chain can be used to describe a sequence of events, the steps in a procedure, or the stages of a process.

When making an events chain, first find the one event that starts the chain. This event is called the initiating event. Then, find the next event and continue until the outcome is reached, as shown in **Figure 4.**

SCIENCE SKILL HANDBOOK F ◆ 187

Science Skill Handbook

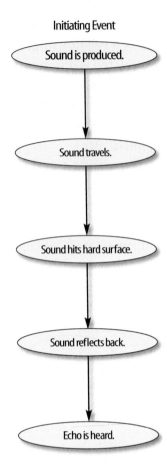

Figure 4 Events-chain concept maps show the order of steps in a process or event. This concept map shows how a sound makes an echo.

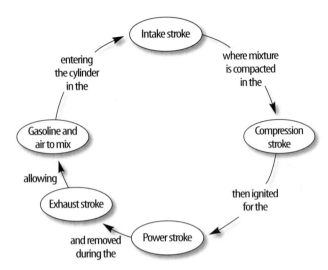

Figure 5 A cycle map shows events that occur in a cycle.

Cycle Map A specific type of events chain is a cycle map. It is used when the series of events do not produce a final outcome, but instead relate back to the beginning event, such as in **Figure 5.** Therefore, the cycle repeats itself.

To make a cycle map, first decide what event is the beginning event. This is also called the initiating event. Then list the next events in the order that they occur, with the last event relating back to the initiating event. Words can be written between the events that describe what happens from one event to the next. The number of events in a cycle map can vary, but usually contain three or more events.

Spider Map A type of concept map that you can use for brainstorming is the spider map. When you have a central idea, you might find that you have a jumble of ideas that relate to it but are not necessarily clearly related to each other. The spider map on sound in **Figure 6** shows that if you write these ideas outside the main concept, then you can begin to separate and group unrelated terms so they become more useful.

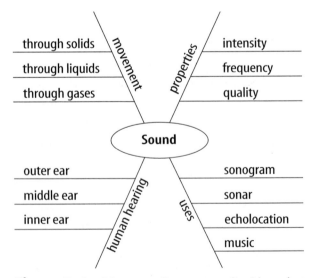

Figure 6 A spider map allows you to list ideas that relate to a central topic but not necessarily to one another.

188 ◆ **F** STUDENT RESOURCES

Science Skill Handbook

Figure 7 This Venn diagram compares and contrasts two substances made from carbon.

Venn Diagram To illustrate how two subjects compare and contrast you can use a Venn diagram. You can see the characteristics that the subjects have in common and those that they do not, shown in **Figure 7.**

To create a Venn diagram, draw two overlapping ovals that that are big enough to write in. List the characteristics unique to one subject in one oval, and the characteristics of the other subject in the other oval. The characteristics in common are listed in the overlapping section.

Make and Use Tables One way to organize information so it is easier to understand is to use a table. Tables can contain numbers, words, or both.

To make a table, list the items to be compared in the first column and the characteristics to be compared in the first row. The title should clearly indicate the content of the table, and the column or row heads should be clear. Notice that in **Table 1** the units are included.

| Table 1 Recyclables Collected During Week ||||
Day of Week	Paper (kg)	Aluminum (kg)	Glass (kg)
Monday	5.0	4.0	12.0
Wednesday	4.0	1.0	10.0
Friday	2.5	2.0	10.0

Make a Model One way to help you better understand the parts of a structure, the way a process works, or to show things too large or small for viewing is to make a model. For example, an atomic model made of a plastic-ball nucleus and pipe-cleaner electron shells can help you visualize how the parts of an atom relate to each other. Other types of models can by devised on a computer or represented by equations.

Form a Hypothesis

A possible explanation based on previous knowledge and observations is called a hypothesis. After researching gasoline types and recalling previous experiences in your family's car you form a hypothesis—our car runs more efficiently because we use premium gasoline. To be valid, a hypothesis has to be something you can test by using an investigation.

Predict When you apply a hypothesis to a specific situation, you predict something about that situation. A prediction makes a statement in advance, based on prior observation, experience, or scientific reasoning. People use predictions to make everyday decisions. Scientists test predictions by performing investigations. Based on previous observations and experiences, you might form a prediction that cars are more efficient with premium gasoline. The prediction can be tested in an investigation.

Design an Experiment A scientist needs to make many decisions before beginning an investigation. Some of these include: how to carry out the investigation, what steps to follow, how to record the data, and how the investigation will answer the question. It also is important to address any safety concerns.

SCIENCE SKILL HANDBOOK F ◆ 189

Science Skill Handbook

Test the Hypothesis

Now that you have formed your hypothesis, you need to test it. Using an investigation, you will make observations and collect data, or information. This data might either support or not support your hypothesis. Scientists collect and organize data as numbers and descriptions.

Follow a Procedure In order to know what materials to use, as well as how and in what order to use them, you must follow a procedure. **Figure 8** shows a procedure you might follow to test your hypothesis.

Procedure
1. Use regular gasoline for two weeks.
2. Record the number of kilometers between fill-ups and the amount of gasoline used.
3. Switch to premium gasoline for two weeks.
4. Record the number of kilometers between fill-ups and the amount of gasoline used.

Figure 8 A procedure tells you what to do step by step.

Identify and Manipulate Variables and Controls In any experiment, it is important to keep everything the same except for the item you are testing. The one factor you change is called the independent variable. The change that results is the dependent variable. Make sure you have only one independent variable, to assure yourself of the cause of the changes you observe in the dependent variable. For example, in your gasoline experiment the type of fuel is the independent variable. The dependent variable is the efficiency.

Many experiments also have a control—an individual instance or experimental subject for which the independent variable is not changed. You can then compare the test results to the control results. To design a control you can have two cars of the same type. The control car uses regular gasoline for four weeks. After you are done with the test, you can compare the experimental results to the control results.

Collect Data

Whether you are carrying out an investigation or a short observational experiment, you will collect data, as shown in **Figure 9**. Scientists collect data as numbers and descriptions and organize it in specific ways.

Observe Scientists observe items and events, then record what they see. When they use only words to describe an observation, it is called qualitative data. Scientists' observations also can describe how much there is of something. These observations use numbers, as well as words, in the description and are called quantitative data. For example, if a sample of the element gold is described as being "shiny and very dense" the data are qualitative. Quantitative data on this sample of gold might include "a mass of 30 g and a density of 19.3 g/cm^3."

Figure 9 Collecting data is one way to gather information directly.

Science Skill Handbook

Figure 10 Record data neatly and clearly so it is easy to understand.

When you make observations you should examine the entire object or situation first, and then look carefully for details. It is important to record observations accurately and completely. Always record your notes immediately as you make them, so you do not miss details or make a mistake when recording results from memory. Never put unidentified observations on scraps of paper. Instead they should be recorded in a notebook, like the one in **Figure 10.** Write your data neatly so you can easily read it later. At each point in the experiment, record your observations and label them. That way, you will not have to determine what the figures mean when you look at your notes later. Set up any tables that you will need to use ahead of time, so you can record any observations right away. Remember to avoid bias when collecting data by not including personal thoughts when you record observations. Record only what you observe.

Estimate Scientific work also involves estimating. To estimate is to make a judgment about the size or the number of something without measuring or counting. This is important when the number or size of an object or population is too large or too difficult to accurately count or measure.

Sample Scientists may use a sample or a portion of the total number as a type of estimation. To sample is to take a small, representative portion of the objects or organisms of a population for research. By making careful observations or manipulating variables within that portion of the group, information is discovered and conclusions are drawn that might apply to the whole population. A poorly chosen sample can be unrepresentative of the whole. If you were trying to determine the rainfall in an area, it would not be best to take a rainfall sample from under a tree.

Measure You use measurements everyday. Scientists also take measurements when collecting data. When taking measurements, it is important to know how to use measuring tools properly. Accuracy also is important.

Length To measure length, the distance between two points, scientists use meters. Smaller measurements might be measured in centimeters or millimeters.

Length is measured using a metric ruler or meter stick. When using a metric ruler, line up the 0-cm mark with the end of the object being measured and read the number of the unit where the object ends. Look at the metric ruler shown in **Figure 11.** The centimeter lines are the long, numbered lines, and the shorter lines are millimeter lines. In this instance, the length would be 4.50 cm.

Figure 11 This metric ruler has centimeter and millimeter divisions.

SCIENCE SKILL HANDBOOK F ◆ 191

Science Skill Handbook

Mass The SI unit for mass is the kilogram (kg). Scientists can measure mass using units formed by adding metric prefixes to the unit gram (g), such as milligram (mg). To measure mass, you might use a triple-beam balance similar to the one shown in **Figure 12.** The balance has a pan on one side and a set of beams on the other side. Each beam has a rider that slides on the beam.

When using a triple-beam balance, place an object on the pan. Slide the largest rider along its beam until the pointer drops below zero. Then move it back one notch. Repeat the process for each rider proceeding from the larger to smaller until the pointer swings an equal distance above and below the zero point. Sum the masses on each beam to find the mass of the object. Move all riders back to zero when finished.

Instead of putting materials directly on the balance, scientists often take a tare of a container. A tare is the mass of a container into which objects or substances are placed for measuring their masses. To mass objects or substances, find the mass of a clean container. Remove the container from the pan, and place the object or substances in the container. Find the mass of the container with the materials in it. Subtract the mass of the empty container from the mass of the filled container to find the mass of the materials you are using.

Figure 12 A triple-beam balance is used to determine the mass of an object.

Figure 13 Graduated cylinders measure liquid volume.

Liquid Volume To measure liquids, the unit used is the liter. When a smaller unit is needed, scientists might use a milliliter. Because a milliliter takes up the volume of a cube measuring 1 cm on each side it also can be called a cubic centimeter (cm^3 = cm × cm × cm).

You can use beakers and graduated cylinders to measure liquid volume. A graduated cylinder, shown in **Figure 13,** is marked from bottom to top in milliliters. In lab, you might use a 10-mL graduated cylinder or a 100-mL graduated cylinder. When measuring liquids, notice that the liquid has a curved surface. Look at the surface at eye level, and measure the bottom of the curve. This is called the meniscus. The graduated cylinder in **Figure 13** contains 79.0 mL, or 79.0 cm^3, of a liquid.

Temperature Scientists often measure temperature using the Celsius scale. Pure water has a freezing point of 0°C and boiling point of 100°C. The unit of measurement is degrees Celsius. Two other scales often used are the Fahrenheit and Kelvin scales.

Science Skill Handbook

Figure 14 A thermometer measures the temperature of an object.

Scientists use a thermometer to measure temperature. Most thermometers in a laboratory are glass tubes with a bulb at the bottom end containing a liquid such as colored alcohol. The liquid rises or falls with a change in temperature. To read a glass thermometer like the thermometer in **Figure 14,** rotate it slowly until a red line appears. Read the temperature where the red line ends.

Form Operational Definitions An operational definition defines an object by how it functions, works, or behaves. For example, when you are playing hide and seek and a tree is home base, you have created an operational definition for a tree.

Objects can have more than one operational definition. For example, a ruler can be defined as a tool that measures the length of an object (how it is used). It can also be a tool with a series of marks used as a standard when measuring (how it works).

Analyze the Data

To determine the meaning of your observations and investigation results, you will need to look for patterns in the data. Then you must think critically to determine what the data mean. Scientists use several approaches when they analyze the data they have collected and recorded. Each approach is useful for identifying specific patterns.

Interpret Data The word *interpret* means "to explain the meaning of something." When analyzing data from an experiement, try to find out what the data show. Identify the control group and the test group to see whether or not changes in the independent variable have had an effect. Look for differences in the dependent variable between the control and test groups.

Classify Sorting objects or events into groups based on common features is called classifying. When classifying, first observe the objects or events to be classified. Then select one feature that is shared by some members in the group, but not by all. Place those members that share that feature in a subgroup. You can classify members into smaller and smaller subgroups based on characteristics. Remember that when you classify, you are grouping objects or events for a purpose. Keep your purpose in mind as you select the features to form groups and subgroups.

Compare and Contrast Observations can be analyzed by noting the similarities and differences between two more objects or events that you observe. When you look at objects or events to see how they are similar, you are comparing them. Contrasting is looking for differences in objects or events.

Science Skill Handbook

Recognize Cause and Effect A cause is a reason for an action or condition. The effect is that action or condition. When two events happen together, it is not necessarily true that one event caused the other. Scientists must design a controlled investigation to recognize the exact cause and effect.

Draw Conclusions

When scientists have analyzed the data they collected, they proceed to draw conclusions about the data. These conclusions are sometimes stated in words similar to the hypothesis that you formed earlier. They may confirm a hypothesis, or lead you to a new hypothesis.

Infer Scientists often make inferences based on their observations. An inference is an attempt to explain observations or to indicate a cause. An inference is not a fact, but a logical conclusion that needs further investigation. For example, you may infer that a fire has caused smoke. Until you investigate, however, you do not know for sure.

Apply When you draw a conclusion, you must apply those conclusions to determine whether the data supports the hypothesis. If your data do not support your hypothesis, it does not mean that the hypothesis is wrong. It means only that the result of the investigation did not support the hypothesis. Maybe the experiment needs to be redesigned, or some of the initial observations on which the hypothesis was based were incomplete or biased. Perhaps more observation or research is needed to refine your hypothesis. A successful investigation does not always come out the way you originally predicted.

Avoid Bias Sometimes a scientific investigation involves making judgments. When you make a judgment, you form an opinion. It is important to be honest and not to allow any expectations of results to bias your judgments. This is important throughout the entire investigation, from researching to collecting data to drawing conclusions.

Communicate

The communication of ideas is an important part of the work of scientists. A discovery that is not reported will not advance the scientific community's understanding or knowledge. Communication among scientists also is important as a way of improving their investigations.

Scientists communicate in many ways, from writing articles in journals and magazines that explain their investigations and experiments, to announcing important discoveries on television and radio. Scientists also share ideas with colleagues on the Internet or present them as lectures, like the student is doing in **Figure 15.**

Figure 15 A student communicates to his peers about his investigation.

Science Skill Handbook

SAFETY SYMBOLS

	HAZARD	EXAMPLES	PRECAUTION	REMEDY
DISPOSAL	Special disposal procedures need to be followed.	certain chemicals, living organisms	Do not dispose of these materials in the sink or trash can.	Dispose of wastes as directed by your teacher.
BIOLOGICAL	Organisms or other biological materials that might be harmful to humans	bacteria, fungi, blood, unpreserved tissues, plant materials	Avoid skin contact with these materials. Wear mask or gloves.	Notify your teacher if you suspect contact with material. Wash hands thoroughly.
EXTREME TEMPERATURE	Objects that can burn skin by being too cold or too hot	boiling liquids, hot plates, dry ice, liquid nitrogen	Use proper protection when handling.	Go to your teacher for first aid.
SHARP OBJECT	Use of tools or glassware that can easily puncture or slice skin	razor blades, pins, scalpels, pointed tools, dissecting probes, broken glass	Practice common-sense behavior and follow guidelines for use of the tool.	Go to your teacher for first aid.
FUME	Possible danger to respiratory tract from fumes	ammonia, acetone, nail polish remover, heated sulfur, moth balls	Make sure there is good ventilation. Never smell fumes directly. Wear a mask.	Leave foul area and notify your teacher immediately.
ELECTRICAL	Possible danger from electrical shock or burn	improper grounding, liquid spills, short circuits, exposed wires	Double-check setup with teacher. Check condition of wires and apparatus.	Do not attempt to fix electrical problems. Notify your teacher immediately.
IRRITANT	Substances that can irritate the skin or mucous membranes of the respiratory tract	pollen, moth balls, steel wool, fiberglass, potassium permanganate	Wear dust mask and gloves. Practice extra care when handling these materials.	Go to your teacher for first aid.
CHEMICAL	Chemicals can react with and destroy tissue and other materials	bleaches such as hydrogen peroxide; acids such as sulfuric acid, hydrochloric acid; bases such as ammonia, sodium hydroxide	Wear goggles, gloves, and an apron.	Immediately flush the affected area with water and notify your teacher.
TOXIC	Substance may be poisonous if touched, inhaled, or swallowed.	mercury, many metal compounds, iodine, poinsettia plant parts	Follow your teacher's instructions.	Always wash hands thoroughly after use. Go to your teacher for first aid.
FLAMMABLE	Flammable chemicals may be ignited by open flame, spark, or exposed heat.	alcohol, kerosene, potassium permanganate	Avoid open flames and heat when using flammable chemicals.	Notify your teacher immediately. Use fire safety equipment if applicable.
OPEN FLAME	Open flame in use, may cause fire.	hair, clothing, paper, synthetic materials	Tie back hair and loose clothing. Follow teacher's instruction on lighting and extinguishing flames.	Notify your teacher immediately. Use fire safety equipment if applicable.

 Eye Safety Proper eye protection should be worn at all times by anyone performing or observing science activities.

 Clothing Protection This symbol appears when substances could stain or burn clothing.

 Animal Safety This symbol appears when safety of animals and students must be ensured.

 Handwashing After the lab, wash hands with soap and water before removing goggles.

Science Skill Handbook

Safety in the Science Laboratory

The science laboratory is a safe place to work if you follow standard safety procedures. Being responsible for your own safety helps to make the entire laboratory a safer place for everyone. When performing any lab, read and apply the caution statements and safety symbol listed at the beginning of the lab.

General Safety Rules

1. Obtain your teacher's permission to begin all investigations and use laboratory equipment.

2. Study the procedure. Ask your teacher any questions. Be sure you understand safety symbols shown on the page.

3. Notify your teacher about allergies or other health conditions which can affect your participation in a lab.

4. Learn and follow use and safety procedures for your equipment. If unsure, ask your teacher.

5. Never eat, drink, chew gum, apply cosmetics, or do any personal grooming in the lab. Never use lab glassware as food or drink containers. Keep your hands away from your face and mouth.

6. Know the location and proper use of the safety shower, eye wash, fire blanket, and fire alarm.

Prevent Accidents

1. Use the safety equipment provided to you. Goggles and a safety apron should be worn during investigations.

2. Do NOT use hair spray, mousse, or other flammable hair products. Tie back long hair and tie down loose clothing.

3. Do NOT wear sandals or other open-toed shoes in the lab.

4. Remove jewelry on hands and wrists. Loose jewelry, such as chains and long necklaces, should be removed to prevent them from getting caught in equipment.

5. Do not taste any substances or draw any material into a tube with your mouth.

6. Proper behavior is expected in the lab. Practical jokes and fooling around can lead to accidents and injury.

7. Keep your work area uncluttered.

Laboratory Work

1. Collect and carry all equipment and materials to your work area before beginning a lab.

2. Remain in your own work area unless given permission by your teacher to leave it.

Science Skill Handbook

3. Dispose of chemicals and other materials as directed by your teacher. Place broken glass and solid substances in the proper containers. Never discard materials in the sink.

4. Clean your work area.

5. Wash your hands with soap and water thoroughly BEFORE removing your goggles.

Emergencies

1. Report any fire, electrical shock, glassware breakage, spill, or injury, no matter how small, to your teacher immediately. Follow his or her instructions.

2. If your clothing should catch fire, STOP, DROP, and ROLL. If possible, smother it with the fire blanket or get under a safety shower. NEVER RUN.

3. If a fire should occur, turn off all gas and leave the room according to established procedures.

4. In most instances, your teacher will clean up spills. Do NOT attempt to clean up spills unless you are given permission and instructions to do so.

5. If chemicals come into contact with your eyes or skin, notify your teacher immediately. Use the eyewash or flush your skin or eyes with large quantities of water.

6. The fire extinguisher and first-aid kit should only be used by your teacher unless it is an extreme emergency and you have been given permission.

7. If someone is injured or becomes ill, only a professional medical provider or someone certified in first aid should perform first-aid procedures.

3. Always slant test tubes away from yourself and others when heating them, adding substances to them, or rinsing them.

4. If instructed to smell a substance in a container, hold the container a short distance away and fan vapors towards your nose.

5. Do NOT substitute other chemicals/substances for those in the materials list unless instructed to do so by your teacher.

6. Do NOT take any materials or chemicals outside of the laboratory.

7. Stay out of storage areas unless instructed to be there and supervised by your teacher.

Laboratory Cleanup

1. Turn off all burners, water, and gas, and disconnect all electrical devices.

2. Clean all pieces of equipment and return all materials to their proper places.

SCIENCE SKILL HANDBOOK F ◆ **197**

Extra Try at Home Labs

EXTRA Try at Home Labs

From Your Kitchen, Junk Drawer, or Yard

1 Panning Minerals

Real-World Question
How can minerals be separated from sand?

Possible Materials
- large, aluminum pie pan
- gallon jug filled with water
- empty gallon jug
- clean sand
- funnel
- coffee filter
- squirt bottle of water
- magnifying lens
- white paper
- hand magnet

Procedure
1. Conduct this lab outdoors.
2. Line the funnel with a coffee filter. Insert the funnel stem into an empty gallon jug.
3. Add a small amount of sand to the pie pan. Add some water and swirl the pan.
4. Continue to shake and swirl the pan until only black sand is left in the pan.
5. Use the squirt bottle to wash the black sand into the coffee filter. Repeat steps 3–5 until you have a good sample of black sand.
6. Let the black sand dry. Then observe it with a magnifying lens. Test the sand with a magnet.

Conclude and Apply
1. Why was black sand left in the gold pan after swirling it?
2. Describe how the sand looked under the lens. Did you see any well-shaped crystals?
3. What happened when you tested the sand with a magnet? Explain.

2 Changing Rocks

Real-World Question
How can the change of metamorphic rock be modeled?

Possible Materials
- soil
- water
- measuring cup
- bowl
- spoon
- shale sample
- slate sample
- schist sample
- gneiss sample

Procedure
1. Mix equal parts of soil and water in a measuring cup or bowl. Stir the mixture until you make mud.
2. Place the bowl of mud on the table near the top edge.
3. Lay a sample of shale below the mud, a sample of slate below the shale, a sample of schist below the slate, a sample of gneiss below the schist.
4. Observe the different stages of sedimentary and metamorphic rocks that are formed by heat and pressure over long periods of time.

Conclude and Apply
1. Identify which rock sample(s) are sedimentary rock and which sample(s) are metamorphic rock.
2. Infer which type of rock is found at the greatest depth beneath the surface of Earth.

Adult supervision required for all labs.

Extra Try at Home Labs

3 Why recycle?

Real-World Question
What are the effects of throwing out aluminum cans instead of recycling them?

Possible Materials
- calculator
- an aluminum can

Procedure
1. An aluminum can has a mass of about 13 g.

2. Convert the can's mass from grams to kilograms by dividing the mass by 1,000.
3. Find the volume of the can (in milliliters) on the label.
4. Convert the volume from milliliters to liters by dividing it by 1,000.

Conclude and Apply
1. Calculate the mass of aluminum cans thrown out by Americans each year by multiplying the mass of the can in kilograms times 50,000,000,000.
2. Calculate the amount of fuel needed to remake the cans thrown out by Americans each year by multiplying the volume of the can in liters times 50,000,000,000 and dividing your total by 2.
3. Infer the environmental effects of throwing out aluminum cans instead of recycling them.

4 Measuring Movement

Real-World Question
How can we model continental drift?

Possible Materials
- flashlight, nail, rubber band or tape, thick circle of paper
- protractor
- mirror
- stick-on notepad paper
- marker
- metric ruler
- calculator

Procedure
1. Cut a circle of paper to fit around the lens of the flashlight. Use a nail to make a hole in the paper. Fasten the paper with the rubber band or tape. You should now have a flashlight that shines a focused beam of light.
2. Direct the light beam of the flashlight on a protractor held horizontally so that the beam lines up to the 90° mark.
3. Darken a room and aim the light beam at a mirror from an angle. Measure the angle. Observe where the reflected beam hits the wall.
4. Have a partner place a stick-on note on the wall and mark the location of the beam on the paper with a marker.
5. Move the flashlight to a 100° angle and mark the beam's location on the wall with a second note.
6. Measure the distance between the two points on the wall and divide by ten to determine the distance per degree.

Conclude and Apply
1. What was the distance per degree of your measurements?
2. Calculate what the distance would be between the first spot and a third spot marking the location of the flashlight at a 40° angle. Test your calculations.
3. Explain how this lab models measuring continental drift.

Adult supervision required for all labs.

Extra Try at Home Labs

5 Making Waves

Real-World Question
What do earthquake waves look like?

Possible Materials
- rope (3-m length)
- coiled spring
- garden hose

Procedure
1. With a partner, stretch a coiled spring out on the floor. Firmly push your side of the spring in and out toward your partner and observe the waves you created.
2. With a partner, stretch the rope out on the floor. Quickly wave your end side to side and observe the waves you created.
3. Stand with a partner, stretch the rope out, and hold it waist high. Quickly move your hand up and down and observe the waves you created.

Conclude and Apply
1. Infer the type of seismic wave you modeled with the coiled spring.
2. Infer the type of seismic waves you modeled with the rope on the floor.
3. Infer the type of seismic waves you modeled with the rope in the air.

6 Mini Eruptions

Real-World Question
How can we model the eruptions of shield and cinder cone volcanoes?

Possible Materials
- tube of toothpaste
- pin
- unopened bottle of carbonated soda
- newspaper
- paper towels

Procedure
1. Lay down newspaper or paper towels.
2. Press down on the back end of a full tube of toothpaste to move all the paste to the front of the tube.
3. Have a partner press a long pin into the center of the tube. Observe what happens to the toothpaste.
4. Go outside and vigorously shake a bottle of carbonated soda for 1 min.
5. Point the bottle away from other people and quickly remove the cap. Observe what happens to the soda.

Conclude and Apply
1. Describe what happened to the toothpaste and soda.
2. Infer how you modeled a shield volcano eruption.
3. Infer how you modeled a cinder cone volcano eruption.

Technology Skill Handbook

Computer Skills

People who study science rely on computers, like the one in **Figure 16,** to record and store data and to analyze results from investigations. Whether you work in a laboratory or just need to write a lab report with tables, good computer skills are a necessity.

Using the computer comes with responsibility. Issues of ownership, security, and privacy can arise. Remember, if you did not author the information you are using, you must provide a source for your information. Also, anything on a computer can be accessed by others. Do not put anything on the computer that you would not want everyone to know. To add more security to your work, use a password.

Use a Word Processing Program

A computer program that allows you to type your information, change it as many times as you need to, and then print it out is called a word processing program. Word processing programs also can be used to make tables.

Figure 16 A computer will make reports neater and more professional looking.

Learn the Skill To start your word processing program, a blank document, sometimes called "Document 1," appears on the screen. To begin, start typing. To create a new document, click the *New* button on the standard tool bar. These tips will help you format the document.

- The program will automatically move to the next line; press *Enter* if you wish to start a new paragraph.
- Symbols, called non-printing characters, can be hidden by clicking the *Show/Hide* button on your toolbar.
- To insert text, move the cursor to the point where you want the insertion to go, click on the mouse once, and type the text.
- To move several lines of text, select the text and click the *Cut* button on your toolbar. Then position your cursor in the location that you want to move the cut text and click *Paste*. If you move to the wrong place, click *Undo*.
- The spell check feature does not catch words that are misspelled to look like other words, like "cold" instead of "gold." Always reread your document to catch all spelling mistakes.
- To learn about other word processing methods, read the user's manual or click on the *Help* button.
- You can integrate databases, graphics, and spreadsheets into documents by copying from another program and pasting it into your document, or by using desktop publishing (DTP). DTP software allows you to put text and graphics together to finish your document with a professional look. This software varies in how it is used and its capabilities.

TECHNOLOGY SKILL HANDBOOK F ◆ 201

Technology Skill Handbook

Use a Database

A collection of facts stored in a computer and sorted into different fields is called a database. A database can be reorganized in any way that suits your needs.

Learn the Skill A computer program that allows you to create your own database is a database management system (DBMS). It allows you to add, delete, or change information. Take time to get to know the features of your database software.

- Determine what facts you would like to include and research to collect your information.
- Determine how you want to organize the information.
- Follow the instructions for your particular DBMS to set up fields. Then enter each item of data in the appropriate field.
- Follow the instructions to sort the information in order of importance.
- Evaluate the information in your database, and add, delete, or change as necessary.

Use the Internet

The Internet is a global network of computers where information is stored and shared. To use the Internet, like the students in **Figure 17,** you need a modem to connect your computer to a phone line and an Internet Service Provider account.

Learn the Skill To access internet sites and information, use a "Web browser," which lets you view and explore pages on the World Wide Web. Each page is its own site, and each site has its own address, called a URL. Once you have found a Web browser, follow these steps for a search (this also is how you search a database).

Figure 17 The Internet allows you to search a global network for a variety of information.

- Be as specific as possible. If you know you want to research "gold," don't type in "elements." Keep narrowing your search until you find what you want.
- Web sites that end in *.com* are commercial Web sites; *.org, .edu,* and *.gov* are non-profit, educational, or government Web sites.
- Electronic encyclopedias, almanacs, indexes, and catalogs will help locate and select relevant information.
- Develop a "home page" with relative ease. When developing a Web site, NEVER post pictures or disclose personal information such as location, names, or phone numbers. Your school or community usually can host your Web site. A basic understanding of HTML (hypertext mark-up language), the language of Web sites, is necessary. Software that creates HTML code is called authoring software, and can be downloaded free from many Web sites. This software allows text and pictures to be arranged as the software is writing the HTML code.

Technology Skill Handbook

Use a Spreadsheet

A spreadsheet, shown in **Figure 18,** can perform mathematical functions with any data arranged in columns and rows. By entering a simple equation into a cell, the program can perform operations in specific cells, rows, or columns.

Learn the Skill Each column (vertical) is assigned a letter, and each row (horizontal) is assigned a number. Each point where a row and column intersect is called a cell, and is labeled according to where it is located—Column A, Row 1 (A1).

- Decide how to organize the data, and enter it in the correct row or column.
- Spreadsheets can use standard formulas or formulas can be customized to calculate cells.
- To make a change, click on a cell to make it activate, and enter the edited data or formula.
- Spreadsheets also can display your results in graphs. Choose the style of graph that best represents the data.

Figure 18 A spreadsheet allows you to perform mathematical operations on your data.

Use Graphics Software

Adding pictures, called graphics, to your documents is one way to make your documents more meaningful and exciting. This software adds, edits, and even constructs graphics. There is a variety of graphics software programs. The tools used for drawing can be a mouse, keyboard, or other specialized devices. Some graphics programs are simple. Others are complicated, called computer-aided design (CAD) software.

Learn the Skill It is important to have an understanding of the graphics software being used before starting. The better the software is understood, the better the results. The graphics can be placed in a word-processing document.

- Clip art can be found on a variety of internet sites, and on CDs. These images can be copied and pasted into your document.
- When beginning, try editing existing drawings, then work up to creating drawings.
- The images are made of tiny rectangles of color called pixels. Each pixel can be altered.
- Digital photography is another way to add images. The photographs in the memory of a digital camera can be downloaded into a computer, then edited and added to the document.
- Graphics software also can allow animation. The software allows drawings to have the appearance of movement by connecting basic drawings automatically. This is called in-betweening, or tweening.
- Remember to save often.

TECHNOLOGY SKILL HANDBOOK F ◆ 203

Technology Skill Handbook

Presentation Skills

Develop Multimedia Presentations

Most presentations are more dynamic if they include diagrams, photographs, videos, or sound recordings, like the one shown in **Figure 19.** A multimedia presentation involves using stereos, overhead projectors, televisions, computers, and more.

Learn the Skill Decide the main points of your presentation, and what types of media would best illustrate those points.

- Make sure you know how to use the equipment you are working with.
- Practice the presentation using the equipment several times.
- Enlist the help of a classmate to push play or turn lights out for you. Be sure to practice your presentation with him or her.
- If possible, set up all of the equipment ahead of time, and make sure everything is working properly.

Figure 19 These students are engaging the audience using a variety of tools.

Computer Presentations

There are many different interactive computer programs that you can use to enhance your presentation. Most computers have a compact disc (CD) drive that can play both CDs and digital video discs (DVDs). Also, there is hardware to connect a regular CD, DVD, or VCR. These tools will enhance your presentation.

Another method of using the computer to aid in your presentation is to develop a slide show using a computer program. This can allow movement of visuals at the presenter's pace, and can allow for visuals to build on one another.

Learn the Skill In order to create multimedia presentations on a computer, you need to have certain tools. These may include traditional graphic tools and drawing programs, animation programs, and authoring systems that tie everything together. Your computer will tell you which tools it supports. The most important step is to learn about the tools that you will be using.

- Often, color and strong images will convey a point better than words alone. Use the best methods available to convey your point.
- As with other presentations, practice many times.
- Practice your presentation with the tools you and any assistants will be using.
- Maintain eye contact with the audience. The purpose of using the computer is not to prompt the presenter, but to help the audience understand the points of the presentation.

Math Skill Handbook

Math Review

Use Fractions

A fraction compares a part to a whole. In the fraction $\frac{2}{3}$, the 2 represents the part and is the numerator. The 3 represents the whole and is the denominator.

Reduce Fractions To reduce a fraction, you must find the largest factor that is common to both the numerator and the denominator, the greatest common factor (GCF). Divide both numbers by the GCF. The fraction has then been reduced, or it is in its simplest form.

Example Twelve of the 20 chemicals in the science lab are in powder form. What fraction of the chemicals used in the lab are in powder form?

Step 1 Write the fraction.
$$\frac{part}{whole} = \frac{12}{20}$$

Step 2 To find the GCF of the numerator and denominator, list all of the factors of each number.
Factors of 12: 1, 2, 3, 4, 6, 12 (the numbers that divide evenly into 12)
Factors of 20: 1, 2, 4, 5, 10, 20 (the numbers that divide evenly into 20)

Step 3 List the common factors.
1, 2, 4.

Step 4 Choose the greatest factor in the list.
The GCF of 12 and 20 is 4.

Step 5 Divide the numerator and denominator by the GCF.
$$\frac{12 \div 4}{20 \div 4} = \frac{3}{5}$$

In the lab, $\frac{3}{5}$ of the chemicals are in powder form.

Practice Problem At an amusement park, 66 of 90 rides have a height restriction. What fraction of the rides, in its simplest form, has a height restriction?

Add and Subtract Fractions To add or subtract fractions with the same denominator, add or subtract the numerators and write the sum or difference over the denominator. After finding the sum or difference, find the simplest form for your fraction.

Example 1 In the forest outside your house, $\frac{1}{8}$ of the animals are rabbits, $\frac{3}{8}$ are squirrels, and the remainder are birds and insects. How many are mammals?

Step 1 Add the numerators.
$$\frac{1}{8} + \frac{3}{8} = \frac{(1+3)}{8} = \frac{4}{8}$$

Step 2 Find the GCF.
$$\frac{4}{8} \text{ (GCF, 4)}$$

Step 3 Divide the numerator and denominator by the GCF.
$$\frac{4}{4} = 1, \frac{8}{4} = 2$$

$\frac{1}{2}$ of the animals are mammals.

Example 2 If $\frac{7}{16}$ of the Earth is covered by freshwater, and $\frac{1}{16}$ of that is in glaciers, how much freshwater is not frozen?

Step 1 Subtract the numerators.
$$\frac{7}{16} - \frac{1}{16} = \frac{(7-1)}{16} = \frac{6}{16}$$

Step 2 Find the GCF.
$$\frac{6}{16} \text{ (GCF, 2)}$$

Step 3 Divide the numerator and denominator by the GCF.
$$\frac{6}{2} = 3, \frac{16}{2} = 8$$

$\frac{3}{8}$ of the freshwater is not frozen.

Practice Problem A bicycle rider is going 15 km/h for $\frac{4}{9}$ of his ride, 10 km/h for $\frac{2}{9}$ of his ride, and 8 km/h for the remainder of the ride. How much of his ride is he going over 8 km/h?

Math Skill Handbook

Unlike Denominators To add or subtract fractions with unlike denominators, first find the least common denominator (LCD). This is the smallest number that is a common multiple of both denominators. Rename each fraction with the LCD, and then add or subtract. Find the simplest form if necessary.

Example 1 A chemist makes a paste that is $\frac{1}{2}$ table salt (NaCl), $\frac{1}{3}$ sugar ($C_6H_{12}O_6$), and the rest water (H_2O). How much of the paste is a solid?

Step 1 Find the LCD of the fractions.
$\frac{1}{2} + \frac{1}{3}$ (LCD, 6)

Step 2 Rename each numerator and each denominator with the LCD.
$1 \times 3 = 3, \quad 2 \times 3 = 6$
$1 \times 2 = 2, \quad 3 \times 2 = 6$

Step 3 Add the numerators.
$\frac{3}{6} + \frac{2}{6} = \frac{(3+2)}{6} = \frac{5}{6}$

$\frac{5}{6}$ of the paste is a solid.

Example 2 The average precipitation in Grand Junction, CO, is $\frac{7}{10}$ inch in November, and $\frac{3}{5}$ inch in December. What is the total average precipitation?

Step 1 Find the LCD of the fractions.
$\frac{7}{10} + \frac{3}{5}$ (LCD, 10)

Step 2 Rename each numerator and each denominator with the LCD.
$7 \times 1 = 7, \quad 10 \times 1 = 10$
$3 \times 2 = 6, \quad 5 \times 2 = 10$

Step 3 Add the numerators.
$\frac{7}{10} + \frac{6}{10} = \frac{(7+6)}{10} = \frac{13}{10}$

$\frac{13}{10}$ inches total precipitation, or $1\frac{3}{10}$ inches.

Practice Problem On an electric bill, about $\frac{1}{8}$ of the energy is from solar energy and about $\frac{1}{10}$ is from wind power. How much of the total bill is from solar energy and wind power combined?

Example 3 In your body, $\frac{7}{10}$ of your muscle contractions are involuntary (cardiac and smooth muscle tissue). Smooth muscle makes $\frac{3}{15}$ of your muscle contractions. How many of your muscle contractions are made by cardiac muscle?

Step 1 Find the LCD of the fractions.
$\frac{7}{10} - \frac{3}{15}$ (LCD, 30)

Step 2 Rename each numerator and each denominator with the LCD.
$7 \times 3 = 21, \quad 10 \times 3 = 30$
$3 \times 2 = 6, \quad 15 \times 2 = 30$

Step 3 Subtract the numerators.
$\frac{21}{30} - \frac{6}{30} = \frac{(21-6)}{30} = \frac{15}{30}$

Step 4 Find the GCF.
$\frac{15}{30}$ (GCF, 15)
$\frac{1}{2}$

$\frac{1}{2}$ of all muscle contractions are cardiac muscle.

Example 4 Tony wants to make cookies that call for $\frac{3}{4}$ of a cup of flour, but he only has $\frac{1}{3}$ of a cup. How much more flour does he need?

Step 1 Find the LCD of the fractions.
$\frac{3}{4} - \frac{1}{3}$ (LCD, 12)

Step 2 Rename each numerator and each denominator with the LCD.
$3 \times 3 = 9, \quad 4 \times 3 = 12$
$1 \times 4 = 4, \quad 3 \times 4 = 12$

Step 3 Subtract the numerators.
$\frac{9}{12} - \frac{4}{12} = \frac{(9-4)}{12} = \frac{5}{12}$

$\frac{5}{12}$ of a cup of flour.

Practice Problem Using the information provided to you in Example 3 above, determine how many muscle contractions are voluntary (skeletal muscle).

Math Skill Handbook

Multiply Fractions To multiply with fractions, multiply the numerators and multiply the denominators. Find the simplest form if necessary.

Example Multiply $\frac{3}{5}$ by $\frac{1}{3}$.

Step 1 Multiply the numerators and denominators.
$$\frac{3}{5} \times \frac{1}{3} = \frac{(3 \times 1)}{(5 \times 3)} = \frac{3}{15}$$

Step 2 Find the GCF.
$\frac{3}{15}$ (GCF, 3)

Step 3 Divide the numerator and denominator by the GCF.
$\frac{3}{3} = 1, \frac{15}{3} = 5$
$\frac{1}{5}$

$\frac{3}{5}$ multiplied by $\frac{1}{3}$ is $\frac{1}{5}$.

Practice Problem Multiply $\frac{3}{14}$ by $\frac{5}{16}$.

Find a Reciprocal Two numbers whose product is 1 are called multiplicative inverses, or reciprocals.

Example Find the reciprocal of $\frac{3}{8}$.

Step 1 Inverse the fraction by putting the denominator on top and the numerator on the bottom.
$\frac{8}{3}$

The reciprocal of $\frac{3}{8}$ is $\frac{8}{3}$.

Practice Problem Find the reciprocal of $\frac{4}{9}$.

Divide Fractions To divide one fraction by another fraction, multiply the dividend by the reciprocal of the divisor. Find the simplest form if necessary.

Example 1 Divide $\frac{1}{9}$ by $\frac{1}{3}$.

Step 1 Find the reciprocal of the divisor.
The reciprocal of $\frac{1}{3}$ is $\frac{3}{1}$.

Step 2 Multiply the dividend by the reciprocal of the divisor.
$$\frac{\frac{1}{9}}{\frac{1}{3}} = \frac{1}{9} \times \frac{3}{1} = \frac{(1 \times 3)}{(9 \times 1)} = \frac{3}{9}$$

Step 3 Find the GCF.
$\frac{3}{9}$ (GCF, 3)

Step 4 Divide the numerator and denominator by the GCF.
$\frac{3}{3} = 1, \frac{9}{3} = 3$
$\frac{1}{3}$

$\frac{1}{9}$ divided by $\frac{1}{3}$ is $\frac{1}{3}$.

Example 2 Divide $\frac{3}{5}$ by $\frac{1}{4}$.

Step 1 Find the reciprocal of the divisor.
The reciprocal of $\frac{1}{4}$ is $\frac{4}{1}$.

Step 2 Multiply the dividend by the reciprocal of the divisor.
$$\frac{\frac{3}{5}}{\frac{1}{4}} = \frac{3}{5} \times \frac{4}{1} = \frac{(3 \times 4)}{(5 \times 1)} = \frac{12}{5}$$

$\frac{3}{5}$ divided by $\frac{1}{4}$ is $\frac{12}{5}$ or $2\frac{2}{5}$.

Practice Problem Divide $\frac{3}{11}$ by $\frac{7}{10}$.

Math Skill Handbook

Use Ratios

When you compare two numbers by division, you are using a ratio. Ratios can be written 3 to 5, 3:5, or $\frac{3}{5}$. Ratios, like fractions, also can be written in simplest form.

Ratios can represent probabilities, also called odds. This is a ratio that compares the number of ways a certain outcome occurs to the number of outcomes. For example, if you flip a coin 100 times, what are the odds that it will come up heads? There are two possible outcomes, heads or tails, so the odds of coming up heads are 50:100. Another way to say this is that 50 out of 100 times the coin will come up heads. In its simplest form, the ratio is 1:2.

Example 1 A chemical solution contains 40 g of salt and 64 g of baking soda. What is the ratio of salt to baking soda as a fraction in simplest form?

Step 1 Write the ratio as a fraction.
$$\frac{\text{salt}}{\text{baking soda}} = \frac{40}{64}$$

Step 2 Express the fraction in simplest form.
The GCF of 40 and 64 is 8.
$$\frac{40}{64} = \frac{40 \div 8}{64 \div 8} = \frac{5}{8}$$

The ratio of salt to baking soda in the sample is 5:8.

Example 2 Sean rolls a 6-sided die 6 times. What are the odds that the side with a 3 will show?

Step 1 Write the ratio as a fraction.
$$\frac{\text{number of sides with a 3}}{\text{number of sides}} = \frac{1}{6}$$

Step 2 Multiply by the number of attempts.
$$\frac{1}{6} \times 6 \text{ attempts} = \frac{6}{6} \text{ attempts} = 1 \text{ attempt}$$

1 attempt out of 6 will show a 3.

Practice Problem Two metal rods measure 100 cm and 144 cm in length. What is the ratio of their lengths in simplest form?

Use Decimals

A fraction with a denominator that is a power of ten can be written as a decimal. For example, 0.27 means $\frac{27}{100}$. The decimal point separates the ones place from the tenths place.

Any fraction can be written as a decimal using division. For example, the fraction $\frac{5}{8}$ can be written as a decimal by dividing 5 by 8. Written as a decimal, it is 0.625.

Add or Subtract Decimals When adding and subtracting decimals, line up the decimal points before carrying out the operation.

Example 1 Find the sum of 47.68 and 7.80.

Step 1 Line up the decimal places when you write the numbers.
```
  47.68
+  7.80
```

Step 2 Add the decimals.
```
  47.68
+  7.80
  -----
  55.48
```

The sum of 47.68 and 7.80 is 55.48.

Example 2 Find the difference of 42.17 and 15.85.

Step 1 Line up the decimal places when you write the number.
```
  42.17
- 15.85
```

Step 2 Subtract the decimals.
```
  42.17
- 15.85
  -----
  26.32
```

The difference of 42.17 and 15.85 is 26.32.

Practice Problem Find the sum of 1.245 and 3.842.

Math Skill Handbook

Multiply Decimals To multiply decimals, multiply the numbers like any other number, ignoring the decimal point. Count the decimal places in each factor. The product will have the same number of decimal places as the sum of the decimal places in the factors.

Example Multiply 2.4 by 5.9.

Step 1 Multiply the factors like two whole numbers.
24 × 59 = 1416

Step 2 Find the sum of the number of decimal places in the factors. Each factor has one decimal place, for a sum of two decimal places.

Step 3 The product will have two decimal places.
14.16

The product of 2.4 and 5.9 is 14.16.

Practice Problem Multiply 4.6 by 2.2.

Divide Decimals When dividing decimals, change the divisor to a whole number. To do this, multiply both the divisor and the dividend by the same power of ten. Then place the decimal point in the quotient directly above the decimal point in the dividend. Then divide as you do with whole numbers.

Example Divide 8.84 by 3.4.

Step 1 Multiply both factors by 10.
3.4 × 10 = 34, 8.84 × 10 = 88.4

Step 2 Divide 88.4 by 34.

```
      2.6
34)88.4
   −68
    204
   −204
      0
```

8.84 divided by 3.4 is 2.6.

Practice Problem Divide 75.6 by 3.6.

Use Proportions

An equation that shows that two ratios are equivalent is a proportion. The ratios $\frac{2}{4}$ and $\frac{5}{10}$ are equivalent, so they can be written as $\frac{2}{4} = \frac{5}{10}$. This equation is a proportion.

When two ratios form a proportion, the cross products are equal. To find the cross products in the proportion $\frac{2}{4} = \frac{5}{10}$, multiply the 2 and the 10, and the 4 and the 5. Therefore 2 × 10 = 4 × 5, or 20 = 20.

Because you know that both proportions are equal, you can use cross products to find a missing term in a proportion. This is known as solving the proportion.

Example The heights of a tree and a pole are proportional to the lengths of their shadows. The tree casts a shadow of 24 m when a 6-m pole casts a shadow of 4 m. What is the height of the tree?

Step 1 Write a proportion.
$$\frac{\text{height of tree}}{\text{height of pole}} = \frac{\text{length of tree's shadow}}{\text{length of pole's shadow}}$$

Step 2 Substitute the known values into the proportion. Let h represent the unknown value, the height of the tree.
$$\frac{h}{6} = \frac{24}{4}$$

Step 3 Find the cross products.
$h \times 4 = 6 \times 24$

Step 4 Simplify the equation.
$4h = 144$

Step 5 Divide each side by 4.
$$\frac{4h}{4} = \frac{144}{4}$$
$h = 36$

The height of the tree is 36 m.

Practice Problem The ratios of the weights of two objects on the Moon and on Earth are in proportion. A rock weighing 3 N on the Moon weighs 18 N on Earth. How much would a rock that weighs 5 N on the Moon weigh on Earth?

Math Skill Handbook

Use Percentages

The word *percent* means "out of one hundred." It is a ratio that compares a number to 100. Suppose you read that 77 percent of the Earth's surface is covered by water. That is the same as reading that the fraction of the Earth's surface covered by water is $\frac{77}{100}$. To express a fraction as a percent, first find the equivalent decimal for the fraction. Then, multiply the decimal by 100 and add the percent symbol.

Example Express $\frac{13}{20}$ as a percent.

Step 1 Find the equivalent decimal for the fraction.

$$\begin{array}{r} 0.65 \\ 20\overline{)13.00} \\ \underline{12\ 0} \\ 1\ 00 \\ \underline{1\ 00} \\ 0 \end{array}$$

Step 2 Rewrite the fraction $\frac{13}{20}$ as 0.65.

Step 3 Multiply 0.65 by 100 and add the % sign.
$0.65 \times 100 = 65 = 65\%$

So, $\frac{13}{20} = 65\%$.

This also can be solved as a proportion.

Example Express $\frac{13}{20}$ as a percent.

Step 1 Write a proportion.
$\frac{13}{20} = \frac{x}{100}$

Step 2 Find the cross products.
$1300 = 20x$

Step 3 Divide each side by 20.
$\frac{1300}{20} = \frac{20x}{20}$
$65\% = x$

Practice Problem In one year, 73 of 365 days were rainy in one city. What percent of the days in that city were rainy?

Solve One-Step Equations

A statement that two things are equal is an equation. For example, $A = B$ is an equation that states that A is equal to B.

An equation is solved when a variable is replaced with a value that makes both sides of the equation equal. To make both sides equal the inverse operation is used. Addition and subtraction are inverses, and multiplication and division are inverses.

Example 1 Solve the equation $x - 10 = 35$.

Step 1 Find the solution by adding 10 to each side of the equation.
$x - 10 = 35$
$x - 10 + 10 = 35 + 10$
$x = 45$

Step 2 Check the solution.
$x - 10 = 35$
$45 - 10 = 35$
$35 = 35$

Both sides of the equation are equal, so $x = 45$.

Example 2 In the formula $a = bc$, find the value of c if $a = 20$ and $b = 2$.

Step 1 Rearrange the formula so the unknown value is by itself on one side of the equation by dividing both sides by b.

$a = bc$
$\frac{a}{b} = \frac{bc}{b}$
$\frac{a}{b} = c$

Step 2 Replace the variables a and b with the values that are given.

$\frac{a}{b} = c$
$\frac{20}{2} = c$
$10 = c$

Step 3 Check the solution.
$a = bc$
$20 = 2 \times 10$
$20 = 20$

Both sides of the equation are equal, so $c = 10$ is the solution when $a = 20$ and $b = 2$.

Practice Problem In the formula $h = gd$, find the value of d if $g = 12.3$ and $h = 17.4$.

Math Skill Handbook

Use Statistics

The branch of mathematics that deals with collecting, analyzing, and presenting data is statistics. In statistics, there are three common ways to summarize data with a single number—the mean, the median, and the mode.

The **mean** of a set of data is the arithmetic average. It is found by adding the numbers in the data set and dividing by the number of items in the set.

The **median** is the middle number in a set of data when the data are arranged in numerical order. If there were an even number of data points, the median would be the mean of the two middle numbers.

The **mode** of a set of data is the number or item that appears most often.

Another number that often is used to describe a set of data is the range. The **range** is the difference between the largest number and the smallest number in a set of data.

A **frequency table** shows how many times each piece of data occurs, usually in a survey. **Table 2** below shows the results of a student survey on favorite color.

Table 2 Student Color Choice		
Color	Tally	Frequency
red	||||	4
blue	ＨＨ	5
black	||	2
green	|||	3
purple	ＨＨ ||	7
yellow	ＨＨ |	6

Based on the frequency table data, which color is the favorite?

Example The speeds (in m/s) for a race car during five different time trials are 39, 37, 44, 36, and 44.

To find the mean:

Step 1 Find the sum of the numbers.
39 + 37 + 44 + 36 + 44 = 200

Step 2 Divide the sum by the number of items, which is 5.
200 ÷ 5 = 40

The mean is 40 m/s.

To find the median:

Step 1 Arrange the measures from least to greatest.
36, 37, 39, 44, 44

Step 2 Determine the middle measure.
36, 37, <u>39</u>, 44, 44

The median is 39 m/s.

To find the mode:

Step 1 Group the numbers that are the same together.
44, 44, 36, 37, 39

Step 2 Determine the number that occurs most in the set.
<u>44, 44</u>, 36, 37, 39

The mode is 44 m/s.

To find the range:

Step 1 Arrange the measures from largest to smallest.
44, 44, 39, 37, 36

Step 2 Determine the largest and smallest measures in the set.
<u>44</u>, 44, 39, 37, <u>36</u>

Step 3 Find the difference between the largest and smallest measures.
44 − 36 = 8

The range is 8 m/s.

Practice Problem Find the mean, median, mode, and range for the data set 8, 4, 12, 8, 11, 14, 16.

Math Skill Handbook

Use Geometry

The branch of mathematics that deals with the measurement, properties, and relationships of points, lines, angles, surfaces, and solids is called geometry.

Perimeter The **perimeter** (P) is the distance around a geometric figure. To find the perimeter of a rectangle, add the length and width and multiply that sum by two, or $2(l + w)$. To find perimeters of irregular figures, add the length of the sides.

Example 1 Find the perimeter of a rectangle that is 3 m long and 5 m wide.

Step 1 You know that the perimeter is 2 times the sum of the width and length.
$P = 2(3 \text{ m} + 5 \text{ m})$

Step 2 Find the sum of the width and length.
$P = 2(8 \text{ m})$

Step 3 Multiply by 2.
$P = 16 \text{ m}$

The perimeter is 16 m.

Example 2 Find the perimeter of a shape with sides measuring 2 cm, 5 cm, 6 cm, 3 cm.

Step 1 You know that the perimeter is the sum of all the sides.
$P = 2 + 5 + 6 + 3$

Step 2 Find the sum of the sides.
$P = 2 + 5 + 6 + 3$
$P = 16$

The perimeter is 16 cm.

Practice Problem Find the perimeter of a rectangle with a length of 18 m and a width of 7 m.

Practice Problem Find the perimeter of a triangle measuring 1.6 cm by 2.4 cm by 2.4 cm.

Area of a Rectangle The **area** (A) is the number of square units needed to cover a surface. To find the area of a rectangle, multiply the length times the width, or $l \times w$. When finding area, the units also are multiplied. Area is given in square units.

Example Find the area of a rectangle with a length of 1 cm and a width of 10 cm.

Step 1 You know that the area is the length multiplied by the width.
$A = (1 \text{ cm} \times 10 \text{ cm})$

Step 2 Multiply the length by the width. Also multiply the units.
$A = 10 \text{ cm}^2$

The area is 10 cm^2.

Practice Problem Find the area of a square whose sides measure 4 m.

Area of a Triangle To find the area of a triangle, use the formula:

$$A = \frac{1}{2}(\text{base} \times \text{height})$$

The base of a triangle can be any of its sides. The height is the perpendicular distance from a base to the opposite endpoint, or vertex.

Example Find the area of a triangle with a base of 18 m and a height of 7 m.

Step 1 You know that the area is $\frac{1}{2}$ the base times the height.
$A = \frac{1}{2}(18 \text{ m} \times 7 \text{ m})$

Step 2 Multiply $\frac{1}{2}$ by the product of 18×7. Multiply the units.
$A = \frac{1}{2}(126 \text{ m}^2)$
$A = 63 \text{ m}^2$

The area is 63 m^2.

Practice Problem Find the area of a triangle with a base of 27 cm and a height of 17 cm.

Math Skill Handbook

Circumference of a Circle The **diameter** (d) of a circle is the distance across the circle through its center, and the **radius** (r) is the distance from the center to any point on the circle. The radius is half of the diameter. The distance around the circle is called the **circumference** (C). The formula for finding the circumference is:

$$C = 2\pi r \text{ or } C = \pi d$$

The circumference divided by the diameter is always equal to 3.1415926... This nonterminating and nonrepeating number is represented by the Greek letter π (pi). An approximation often used for π is 3.14.

Example 1 Find the circumference of a circle with a radius of 3 m.

Step 1 You know the formula for the circumference is 2 times the radius times π.
$C = 2\pi(3)$

Step 2 Multiply 2 times the radius.
$C = 6\pi$

Step 3 Multiply by π.
$C = 19$ m

The circumference is 19 m.

Example 2 Find the circumference of a circle with a diameter of 24.0 cm.

Step 1 You know the formula for the circumference is the diameter times π.
$C = \pi(24.0)$

Step 2 Multiply the diameter by π.
$C = 75.4$ cm

The circumference is 75.4 cm.

Practice Problem Find the circumference of a circle with a radius of 19 cm.

Area of a Circle The formula for the area of a circle is:
$A = \pi r^2$

Example 1 Find the area of a circle with a radius of 4.0 cm.

Step 1 $A = \pi(4.0)^2$

Step 2 Find the square of the radius.
$A = 16\pi$

Step 3 Multiply the square of the radius by π.
$A = 50$ cm^2

The area of the circle is 50 cm^2.

Example 2 Find the area of a circle with a radius of 225 m.

Step 1 $A = \pi(225)^2$

Step 2 Find the square of the radius.
$A = 50625\pi$

Step 3 Multiply the square of the radius by π.
$A = 158962.5$

The area of the circle is 158,962 m^2.

Example 3 Find the area of a circle whose diameter is 20.0 mm.

Step 1 You know the formula for the area of a circle is the square of the radius times π, and that the radius is half of the diameter.
$A = \pi\left(\frac{20.0}{2}\right)^2$

Step 2 Find the radius.
$A = \pi(10.0)^2$

Step 3 Find the square of the radius.
$A = 100\pi$

Step 4 Multiply the square of the radius by π.
$A = 314$ mm^2

The area is 314 mm^2.

Practice Problem Find the area of a circle with a radius of 16 m.

Math Skill Handbook

Volume The measure of space occupied by a solid is the **volume** (V). To find the volume of a rectangular solid multiply the length times width times height, or $V = l \times w \times h$. It is measured in cubic units, such as cubic centimeters (cm^3).

Example Find the volume of a rectangular solid with a length of 2.0 m, a width of 4.0 m, and a height of 3.0 m.

Step 1 You know the formula for volume is the length times the width times the height.
$V = 2.0 \text{ m} \times 4.0 \text{ m} \times 3.0 \text{ m}$

Step 2 Multiply the length times the width times the height.
$V = 24 \text{ m}^3$

The volume is 24 m^3.

Practice Problem Find the volume of a rectangular solid that is 8 m long, 4 m wide, and 4 m high.

To find the volume of other solids, multiply the area of the base times the height.

Example 1 Find the volume of a solid that has a triangular base with a length of 8.0 m and a height of 7.0 m. The height of the entire solid is 15.0 m.

Step 1 You know that the base is a triangle, and the area of a triangle is $\frac{1}{2}$ the base times the height, and the volume is the area of the base times the height.
$V = \left[\frac{1}{2} (b \times h)\right] \times 15$

Step 2 Find the area of the base.
$V = \left[\frac{1}{2} (8 \times 7)\right] \times 15$
$V = \left(\frac{1}{2} \times 56\right) \times 15$

Step 3 Multiply the area of the base by the height of the solid.
$V = 28 \times 15$
$V = 420 \text{ m}^3$

The volume is 420 m^3.

Example 2 Find the volume of a cylinder that has a base with a radius of 12.0 cm, and a height of 21.0 cm.

Step 1 You know that the base is a circle, and the area of a circle is the square of the radius times π, and the volume is the area of the base times the height.
$V = (\pi r^2) \times 21$
$V = (\pi 12^2) \times 21$

Step 2 Find the area of the base.
$V = 144\pi \times 21$
$V = 452 \times 21$

Step 3 Multiply the area of the base by the height of the solid.
$V = 9490 \text{ cm}^3$

The volume is 9490 cm^3.

Example 3 Find the volume of a cylinder that has a diameter of 15 mm and a height of 4.8 mm.

Step 1 You know that the base is a circle with an area equal to the square of the radius times π. The radius is one-half the diameter. The volume is the area of the base times the height.
$V = (\pi r^2) \times 4.8$
$V = \left[\pi \left(\frac{1}{2} \times 15\right)^2\right] \times 4.8$
$V = (\pi 7.5^2) \times 4.8$

Step 2 Find the area of the base.
$V = 56.25\pi \times 4.8$
$V = 176.63 \times 4.8$

Step 3 Multiply the area of the base by the height of the solid.
$V = 847.8$

The volume is 847.8 mm^3.

Practice Problem Find the volume of a cylinder with a diameter of 7 cm in the base and a height of 16 cm.

Math Skill Handbook

Science Applications

Measure in SI

The metric system of measurement was developed in 1795. A modern form of the metric system, called the International System (SI), was adopted in 1960 and provides the standard measurements that all scientists around the world can understand.

The SI system is convenient because unit sizes vary by powers of 10. Prefixes are used to name units. Look at **Table 3** for some common SI prefixes and their meanings.

Table 3 Some SI Prefixes			
Prefix	Symbol	Meaning	
kilo-	k	1,000	thousand
hecto-	h	100	hundred
deka-	da	10	ten
deci-	d	0.1	tenth
centi-	c	0.01	hundredth
milli-	m	0.001	thousandth

Example How many grams equal one kilogram?

Step 1 Find the prefix *kilo* in **Table 3**.

Step 2 Using **Table 3**, determine the meaning of *kilo*. According to the table, it means 1,000. When the prefix *kilo* is added to a unit, it means that there are 1,000 of the units in a "*kilo*unit."

Step 3 Apply the prefix to the units in the question. The units in the question are grams. There are 1,000 grams in a kilogram.

Practice Problem Is a milligram larger or smaller than a gram? How many of the smaller units equal one larger unit? What fraction of the larger unit does one smaller unit represent?

Dimensional Analysis

Convert SI Units In science, quantities such as length, mass, and time sometimes are measured using different units. A process called dimensional analysis can be used to change one unit of measure to another. This process involves multiplying your starting quantity and units by one or more conversion factors. A conversion factor is a ratio equal to one and can be made from any two equal quantities with different units. If 1,000 mL equal 1 L then two ratios can be made.

$$\frac{1,000 \text{ mL}}{1 \text{ L}} = \frac{1 \text{ L}}{1,000 \text{ mL}} = 1$$

One can covert between units in the SI system by using the equivalents in **Table 3** to make conversion factors.

Example 1 How many cm are in 4 m?

Step 1 Write conversion factors for the units given. From **Table 3**, you know that 100 cm = 1 m. The conversion factors are

$$\frac{100 \text{ cm}}{1 \text{ m}} \text{ and } \frac{1 \text{ m}}{100 \text{ cm}}$$

Step 2 Decide which conversion factor to use. Select the factor that has the units you are converting from (m) in the denominator and the units you are converting to (cm) in the numerator.

$$\frac{100 \text{ cm}}{1 \text{ m}}$$

Step 3 Multiply the starting quantity and units by the conversion factor. Cancel the starting units with the units in the denominator. There are 400 cm in 4 m.

$$4 \text{ m} \times \frac{100 \text{ cm}}{1 \text{ m}} = 400 \text{ cm}$$

Practice Problem How many milligrams are in one kilogram? (Hint: You will need to use two conversion factors from **Table 3**.)

MATH SKILL HANDBOOK F ◆ 215

Math Skill Handbook

Table 4 Unit System Equivalents	
Type of Measurement	Equivalent
Length	1 in = 2.54 cm
	1 yd = 0.91 m
	1 mi = 1.61 km
Mass and Weight*	1 oz = 28.35 g
	1 lb = 0.45 kg
	1 ton (short) = 0.91 tonnes (metric tons)
	1 lb = 4.45 N
Volume	1 in^3 = 16.39 cm^3
	1 qt = 0.95 L
	1 gal = 3.78 L
Area	1 in^2 = 6.45 cm^2
	1 yd^2 = 0.83 m^2
	1 mi^2 = 2.59 km^2
	1 acre = 0.40 hectares
Temperature	°C = $\frac{(°F - 32)}{1.8}$
	K = °C + 273

*Weight is measured in standard Earth gravity.

Convert Between Unit Systems Table 4 gives a list of equivalents that can be used to convert between English and SI units.

Example If a meterstick has a length of 100 cm, how long is the meterstick in inches?

Step 1 Write the conversion factors for the units given. From **Table 4,** 1 in = 2.54 cm.

$$\frac{1 \text{ in}}{2.54 \text{ cm}} \quad \text{and} \quad \frac{2.54 \text{ cm}}{1 \text{ in}}$$

Step 2 Determine which conversion factor to use. You are converting from cm to in. Use the conversion factor with cm on the bottom.

$$\frac{1 \text{ in}}{2.54 \text{ cm}}$$

Step 3 Multiply the starting quantity and units by the conversion factor. Cancel the starting units with the units in the denominator. Round your answer based on the number of significant figures in the conversion factor.

$$100 \text{ cm} \times \frac{1 \text{ in}}{2.54 \text{ cm}} = 39.37 \text{ in}$$

The meterstick is 39.4 in long.

Practice Problem A book has a mass of 5 lbs. What is the mass of the book in kg?

Practice Problem Use the equivalent for in and cm (1 in = 2.54 cm) to show how 1 in^3 = 16.39 cm^3.

Math Skill Handbook

Precision and Significant Digits

When you make a measurement, the value you record depends on the precision of the measuring instrument. This precision is represented by the number of significant digits recorded in the measurement. When counting the number of significant digits, all digits are counted except zeros at the end of a number with no decimal point such as 2,050, and zeros at the beginning of a decimal such as 0.03020. When adding or subtracting numbers with different precision, round the answer to the smallest number of decimal places of any number in the sum or difference. When multiplying or dividing, the answer is rounded to the smallest number of significant digits of any number being multiplied or divided.

Example The lengths 5.28 and 5.2 are measured in meters. Find the sum of these lengths and record your answer using the correct number of significant digits.

Step 1 Find the sum.

$\quad\quad$ 5.28 m $\quad\quad$ 2 digits after the decimal
$\quad$ + 5.2 m $\quad\quad$ 1 digit after the decimal
$\quad\quad$ 10.48 m

Step 2 Round to one digit after the decimal because the least number of digits after the decimal of the numbers being added is 1.

The sum is 10.5 m.

Practice Problem How many significant digits are in the measurement 7,071,301 m? How many significant digits are in the measurement 0.003010 g?

Practice Problem Multiply 5.28 and 5.2 using the rule for multiplying and dividing. Record the answer using the correct number of significant digits.

Scientific Notation

Many times numbers used in science are very small or very large. Because these numbers are difficult to work with scientists use scientific notation. To write numbers in scientific notation, move the decimal point until only one non-zero digit remains on the left. Then count the number of places you moved the decimal point and use that number as a power of ten. For example, the average distance from the Sun to Mars is 227,800,000,000 m. In scientific notation, this distance is 2.278×10^{11} m. Because you moved the decimal point to the left, the number is a positive power of ten.

The mass of an electron is about 0.000 000 000 000 000 000 000 000 000 000 911 kg. Expressed in scientific notation, this mass is 9.11×10^{-31} kg. Because the decimal point was moved to the right, the number is a negative power of ten.

Example Earth is 149,600,000 km from the Sun. Express this in scientific notation.

Step 1 Move the decimal point until one non-zero digit remains on the left.
$\quad\quad$ 1.496 000 00

Step 2 Count the number of decimal places you have moved. In this case, eight.

Step 3 Show that number as a power of ten, 10^8.

The Earth is 1.496×10^8 km from the Sun.

Practice Problem How many significant digits are in 149,600,000 km? How many significant digits are in 1.496×10^8 km?

Practice Problem Parts used in a high performance car must be measured to 7×10^{-6} m. Express this number as a decimal.

Practice Problem A CD is spinning at 539 revolutions per minute. Express this number in scientific notation.

Math Skill Handbook

Make and Use Graphs

Data in tables can be displayed in a graph—a visual representation of data. Common graph types include line graphs, bar graphs, and circle graphs.

Line Graph A line graph shows a relationship between two variables that change continuously. The independent variable is changed and is plotted on the *x*-axis. The dependent variable is observed, and is plotted on the *y*-axis.

Example Draw a line graph of the data below from a cyclist in a long-distance race.

Table 5 Bicycle Race Data	
Time (h)	Distance (km)
0	0
1	8
2	16
3	24
4	32
5	40

Step 1 Determine the *x*-axis and *y*-axis variables. Time varies independently of distance and is plotted on the *x*-axis. Distance is dependent on time and is plotted on the *y*-axis.

Step 2 Determine the scale of each axis. The *x*-axis data ranges from 0 to 5. The *y*-axis data ranges from 0 to 40.

Step 3 Using graph paper, draw and label the axes. Include units in the labels.

Step 4 Draw a point at the intersection of the time value on the *x*-axis and corresponding distance value on the *y*-axis. Connect the points and label the graph with a title, as shown in **Figure 20**.

Figure 20 This line graph shows the relationship between distance and time during a bicycle ride.

Practice Problem A puppy's shoulder height is measured during the first year of her life. The following measurements were collected: (3 mo, 52 cm), (6 mo, 72 cm), (9 mo, 83 cm), (12 mo, 86 cm). Graph this data.

Find a Slope The slope of a straight line is the ratio of the vertical change, rise, to the horizontal change, run.

$$\text{Slope} = \frac{\text{vertical change (rise)}}{\text{horizontal change (run)}} = \frac{\text{change in } y}{\text{change in } x}$$

Example Find the slope of the graph in **Figure 20**.

Step 1 You know that the slope is the change in *y* divided by the change in *x*.

$$\text{Slope} = \frac{\text{change in } y}{\text{change in } x}$$

Step 2 Determine the data points you will be using. For a straight line, choose the two sets of points that are the farthest apart.

$$\text{Slope} = \frac{(40-0) \text{ km}}{(5-0) \text{ hr}}$$

Step 3 Find the change in *y* and *x*.

$$\text{Slope} = \frac{40 \text{ km}}{5 \text{ h}}$$

Step 4 Divide the change in *y* by the change in *x*.

$$\text{Slope} = \frac{8 \text{ km}}{\text{h}}$$

The slope of the graph is 8 km/h.

Math Skill Handbook

Bar Graph To compare data that does not change continuously you might choose a bar graph. A bar graph uses bars to show the relationships between variables. The *x*-axis variable is divided into parts. The parts can be numbers such as years, or a category such as a type of animal. The *y*-axis is a number and increases continuously along the axis.

Example A recycling center collects 4.0 kg of aluminum on Monday, 1.0 kg on Wednesday, and 2.0 kg on Friday. Create a bar graph of this data.

Step 1 Select the *x*-axis and *y*-axis variables. The measured numbers (the masses of aluminum) should be placed on the *y*-axis. The variable divided into parts (collection days) is placed on the *x*-axis.

Step 2 Create a graph grid like you would for a line graph. Include labels and units.

Step 3 For each measured number, draw a vertical bar above the *x*-axis value up to the *y*-axis value. For the first data point, draw a vertical bar above Monday up to 4.0 kg.

Practice Problem Draw a bar graph of the gases in air: 78% nitrogen, 21% oxygen, 1% other gases.

Circle Graph To display data as parts of a whole, you might use a circle graph. A circle graph is a circle divided into sections that represent the relative size of each piece of data. The entire circle represents 100%, half represents 50%, and so on.

Example Air is made up of 78% nitrogen, 21% oxygen, and 1% other gases. Display the composition of air in a circle graph.

Step 1 Multiply each percent by 360° and divide by 100 to find the angle of each section in the circle.

$$78\% \times \frac{360°}{100} = 280.8°$$

$$21\% \times \frac{360°}{100} = 75.6°$$

$$1\% \times \frac{360°}{100} = 3.6°$$

Step 2 Use a compass to draw a circle and to mark the center of the circle. Draw a straight line from the center to the edge of the circle.

Step 3 Use a protractor and the angles you calculated to divide the circle into parts. Place the center of the protractor over the center of the circle and line the base of the protractor over the straight line.

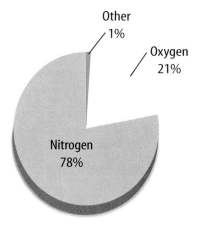

Practice Problem Draw a circle graph to represent the amount of aluminum collected during the week shown in the bar graph to the left.

Weather Map Symbols

Sample Station Model

Sample Plotted Report at Each Station

Precipitation		Wind Speed and Direction		Sky Coverage		Some Types of High Clouds	
≡	Fog	○	0 calm	○	No cover	⌒⊃	Scattered cirrus
★	Snow	/	1–2 knots	⊙	1/10 or less	⊃⊃	Dense cirrus in patches
●	Rain	⌐	3–7 knots	◔	2/10 to 3/10	⊃⌐	Veil of cirrus covering entire sky
⊤⌐	Thunderstorm	⌐⌐	8–12 knots	◑	4/10	⌐	Cirrus not covering entire sky
,	Drizzle	⌐⌐⌐	13–17 knots	◐	–		
∇	Showers	⌐⌐⌐⌐	18–22 knots	◕	6/10		
		⌐⌐⌐⌐⌐	23–27 knots	◕	7/10		
		▼	48–52 knots	◐	Overcast with openings		
		1 knot = 1.852 km/h		●	Completely overcast		

Some Types of Middle Clouds		Some Types of Low Clouds		Fronts and Pressure Systems	
∠	Thin altostratus layer	⌒	Cumulus of fair weather	(H) or High (L) or Low	Center of high- or low-pressure system
∠∠	Thick altostratus layer	⌣	Stratocumulus	▲▲▲▲	Cold front
∠	Thin altostratus in patches	-----	Fractocumulus of bad weather	●●●●	Warm front
∠	Thin altostratus in bands	—	Stratus of fair weather	▲●▲●	Occluded front
				●▼●▼	Stationary front

Rocks

Rocks		
Rock Type	**Rock Name**	**Characteristics**
Igneous (intrusive)	Granite	Large mineral grains of quartz, feldspar, hornblende, and mica. Usually light in color.
	Diorite	Large mineral grains of feldspar, hornblende, and mica. Less quartz than granite. Intermediate in color.
	Gabbro	Large mineral grains of feldspar, augite, and olivine. No quartz. Dark in color.
Igneous (extrusive)	Rhyolite	Small mineral grains of quartz, feldspar, hornblende, and mica, or no visible grains. Light in color.
	Andesite	Small mineral grains of feldspar, hornblende, and mica or no visible grains. Intermediate in color.
	Basalt	Small mineral grains of feldspar, augite, and possibly olivine or no visible grains. No quartz. Dark in color.
	Obsidian	Glassy texture. No visible grains. Volcanic glass. Fracture looks like broken glass.
	Pumice	Frothy texture. Floats in water. Usually light in color.
Sedimentary (detrital)	Conglomerate	Coarse grained. Gravel or pebble-size grains.
	Sandstone	Sand-sized grains 1/16 to 2 mm.
	Siltstone	Grains are smaller than sand but larger than clay.
	Shale	Smallest grains. Often dark in color. Usually platy.
Sedimentary (chemical or organic)	Limestone	Major mineral is calcite. Usually forms in oceans and lakes. Often contains fossils.
	Coal	Forms in swampy areas. Compacted layers of organic material, mainly plant remains.
Sedimentary (chemical)	Rock Salt	Commonly forms by the evaporation of seawater.
Metamorphic (foliated)	Gneiss	Banding due to alternate layers of different minerals, of different colors. Parent rock often is granite.
	Schist	Parallel arrangement of sheetlike minerals, mainly micas. Forms from different parent rocks.
	Phyllite	Shiny or silky appearance. May look wrinkled. Common parent rocks are shale and slate.
	Slate	Harder, denser, and shinier than shale. Common parent rock is shale.
Metamorphic (nonfoliated)	Marble	Calcite or dolomite. Common parent rock is limestone.
	Soapstone	Mainly of talc. Soft with greasy feel.
	Quartzite	Hard with interlocking quartz crystals. Common parent rock is sandstone.

Minerals

Minerals

Mineral (formula)	Color	Streak	Hardness	Breakage Pattern	Uses and Other Properties
Graphite (C)	black to gray	black to gray	1–1.5	basal cleavage (scales)	pencil lead, lubricants for locks, rods to control some small nuclear reactions, battery poles
Galena (PbS)	gray	gray to black	2.5	cubic cleavage perfect	source of lead, used for pipes, shields for X rays, fishing equipment sinkers
Hematite (Fe_2O_3)	black or reddish-brown	reddish-brown	5.5–6.5	irregular fracture	source of iron; converted to pig iron, made into steel
Magnetite (Fe_3O_4)	black	black	6	conchoidal fracture	source of iron, attracts a magnet
Pyrite (FeS_2)	light, brassy, yellow	greenish-black	6–6.5	uneven fracture	fool's gold
Talc ($Mg_3 Si_4O_{10} (OH)_2$)	white, greenish	white	1	cleavage in one direction	used for talcum powder, sculptures, paper, and tabletops
Gypsum ($CaSO_4 \cdot 2H_2O$)	colorless, gray, white, brown	white	2	basal cleavage	used in plaster of paris and dry wall for building construction
Sphalerite (ZnS)	brown, reddish-brown, greenish	light to dark brown	3.5–4	cleavage in six directions	main ore of zinc; used in paints, dyes, and medicine
Muscovite ($KAl_3Si_3 O_{10}(OH)_2$)	white, light gray, yellow, rose, green	colorless	2–2.5	basal cleavage	occurs in large, flexible plates; used as an insulator in electrical equipment, lubricant
Biotite ($K(Mg,Fe)_3 (AlSi_3O_{10}) (OH)_2$)	black to dark brown	colorless	2.5–3	basal cleavage	occurs in large, flexible plates
Halite (NaCl)	colorless, red, white, blue	colorless	2.5	cubic cleavage	salt; soluble in water; a preservative

Minerals

Mineral (formula)	Color	Streak	Hardness	Breakage Pattern	Uses and Other Properties
Calcite ($CaCO_3$)	colorless, white, pale blue	colorless, white	3	cleavage in three directions	fizzes when HCl is added; used in cements and other building materials
Dolomite ($CaMg(CO_3)_2$)	colorless, white, pink, green, gray, black	white	3.5–4	cleavage in three directions	concrete and cement; used as an ornamental building stone
Fluorite (CaF_2)	colorless, white, blue, green, red, yellow, purple	colorless	4	cleavage in four directions	used in the manufacture of optical equipment; glows under ultraviolet light
Hornblende ($(CaNa)_{2-3}(Mg,Al,Fe)_5-(Al,Si)_2Si_6O_{22}(OH)_2$)	green to black	gray to white	5–6	cleavage in two directions	will transmit light on thin edges; 6-sided cross section
Feldspar ($KAlSi_3O_8$) ($NaAlSi_3O_8$), ($CaAl_2Si_2O_8$)	colorless, white to gray, green	colorless	6	two cleavage planes meet at 90° angle	used in the manufacture of ceramics
Augite ($((Ca,Na)(Mg,Fe,Al)(Al,Si)_2O_6$)	black	colorless	6	cleavage in two directions	square or 8-sided cross section
Olivine ($(Mg,Fe)_2SiO_4$)	olive, green	none	6.5–7	conchoidal fracture	gemstones, refractory sand
Quartz (SiO_2)	colorless, various colors	none	7	conchoidal fracture	used in glass manufacture, electronic equipment, radios, computers, watches, gemstones

Reference Handbooks

Reference Handbooks

Metal
Metalloid
Nonmetal

The color of an element's block tells you if the element is a metal, nonmetal, or metalloid.

Science Online
Visit bookf.msscience.com for updates to the periodic table.

13	14	15	16	17	18
					Helium 2 He 4.003
Boron 5 B 10.811	Carbon 6 C 12.011	Nitrogen 7 N 14.007	Oxygen 8 O 15.999	Fluorine 9 F 18.998	Neon 10 Ne 20.180
Aluminum 13 Al 26.982	Silicon 14 Si 28.086	Phosphorus 15 P 30.974	Sulfur 16 S 32.065	Chlorine 17 Cl 35.453	Argon 18 Ar 39.948

10	11	12						
Nickel 28 Ni 58.693	Copper 29 Cu 63.546	Zinc 30 Zn 65.409	Gallium 31 Ga 69.723	Germanium 32 Ge 72.64	Arsenic 33 As 74.922	Selenium 34 Se 78.96	Bromine 35 Br 79.904	Krypton 36 Kr 83.798
Palladium 46 Pd 106.42	Silver 47 Ag 107.868	Cadmium 48 Cd 112.411	Indium 49 In 114.818	Tin 50 Sn 118.710	Antimony 51 Sb 121.760	Tellurium 52 Te 127.60	Iodine 53 I 126.904	Xenon 54 Xe 131.293
Platinum 78 Pt 195.078	Gold 79 Au 196.967	Mercury 80 Hg 200.59	Thallium 81 Tl 204.383	Lead 82 Pb 207.2	Bismuth 83 Bi 208.980	Polonium 84 Po (209)	Astatine 85 At (210)	Radon 86 Rn (222)
Darmstadtium 110 Ds (281)	Roentgenium 111 Rg (272)	Ununbium * 112 Uub (285)		Ununquadium * 114 Uuq (289)				

* The names and symbols for elements 112 and 114 are temporary. Final names will be selected when the elements' discoveries are verified.

Europium 63 Eu 151.964	Gadolinium 64 Gd 157.25	Terbium 65 Tb 158.925	Dysprosium 66 Dy 162.500	Holmium 67 Ho 164.930	Erbium 68 Er 167.259	Thulium 69 Tm 168.934	Ytterbium 70 Yb 173.04	Lutetium 71 Lu 174.967
Americium 95 Am (243)	Curium 96 Cm (247)	Berkelium 97 Bk (247)	Californium 98 Cf (251)	Einsteinium 99 Es (252)	Fermium 100 Fm (257)	Mendelevium 101 Md (258)	Nobelium 102 No (259)	Lawrencium 103 Lr (262)

Topographic Map Symbols

Topographic Map Symbols

————	Primary highway, hard surface	⁓	Index contour
═══▭═══	Secondary highway, hard surface	⋯⋯⋯	Supplementary contour
════════	Light-duty road, hard or improved surface	⁓	Intermediate contour
≡≡≡≡≡≡≡≡	Unimproved road	⌒⌒	Depression contours
┼┼┼┼┼┼	Railroad: single track		
╪╪╪╪╪╪	Railroad: multiple track	— — —	Boundaries: national
╪┼╪┼╪┼	Railroads in juxtaposition	——— —	State
		——— — ⋯	County, parish, municipal
▪▮▬▨	Buildings	——— —	Civil township, precinct, town, barrio
⋰⊹ cem	Schools, church, and cemetery	———⋅——	Incorporated city, village, town, hamlet
▫◻▨	Buildings (barn, warehouse, etc.)	⋅—⋅—⋅⋅	Reservation, national or state
○ ○	Wells other than water (labeled as to type)	----------	Small park, cemetery, airport, etc.
•••⊘	Tanks: oil, water, etc. (labeled only if water)	—⋅⋅—⋅⋅—	Land grant
⊙ ⋎	Located or landmark object; windmill	———————	Township or range line, U.S. land survey
✕ ×	Open pit, mine, or quarry; prospect	— — — —	Township or range line, approximate location
≋≋	Marsh (swamp)		
≋≋	Wooded marsh	∿∿	Perennial streams
▢	Woods or brushwood	→—←	Elevated aqueduct
⋮⋮⋮	Vineyard	○ ∽	Water well and spring
░░	Land subject to controlled inundation	⌇	Small rapids
≡≡	Submerged marsh	░	Large rapids
▨	Mangrove	≣≣≣	Intermittent lake
⋮⋮⋮	Orchard	≈≈	Intermittent stream
⋰⋰	Scrub	→═══←	Aqueduct tunnel
▓	Urban area	░░	Glacier
		⌇	Small falls
x7369	Spot elevation	▬	Large falls
670	Water elevation	░	Dry lake bed

Glossary/Glosario

Cómo usar el glosario en español:
1. Busca el término en inglés que desees encontrar.
2. El término en español, junto con la definición, se encuentran en la columna de la derecha.

Pronunciation Key

Use the following key to help you sound out words in the glossary.

a............	back (BAK)	ew............	food (FEWD)
ay...........	day (DAY)	yoo...........	pure (PYOOR)
ah...........	father (FAH thur)	yew...........	few (FYEW)
ow...........	flower (FLOW ur)	uh............	comma (CAH muh)
ar...........	car (CAR)	u (+ con).....	rub (RUB)
e............	less (LES)	sh............	shelf (SHELF)
ee...........	leaf (LEEF)	ch............	nature (NAY chur)
ih...........	trip (TRIHP)	g.............	gift (GIHFT)
i (i + con + e)..	idea (i DEE uh)	j.............	gem (JEM)
oh...........	go (GOH)	ing...........	sing (SING)
aw...........	soft (SAWFT)	zh............	vision (VIH zhun)
or...........	orbit (OR buht)	k.............	cake (KAYK)
oy...........	coin (COYN)	s.............	seed, cent (SEED, SENT)
oo...........	foot (FOOT)	z.............	zone, raise (ZOHN, RAYZ)

English / Español

A

asthenosphere (as THE nuh sfihr): plasticlike layer of Earth on which the lithospheric plates float and move around. (p. 106)

astenosfera: capa flexible de la Tierra en la que las placas litosféricas flotan y se mueven de un lugar a otro. (p. 106)

B

basaltic: describes dense, dark-colored igneous rock formed from magma rich in magnesium and iron and poor in silica. (p. 43)

basáltica: roca ígnea densa de color oscuro que se forma a partir de magma rico en magnesio y hierro pero pobre en sílice. (p. 43)

batholith: largest intrusive igneous rock body that forms when magma being forced upward toward Earth's crust cools slowly and solidifies underground. (p. 172)

batolito: gran cuerpo rocoso ígneo intrusivo que se forma cuando el magma es forzado a salir a la superficie de la corteza terrestre, se enfría lentamente y se solidifica en el subsuelo. (p. 172)

biomass energy: renewable energy derived from burning organic materials such as wood and alcohol. (p. 79)

energía de biomasa: energía renovable derivada de la combustión de materiales orgánicos tales como la madera y el alcohol. (p. 79)

C

caldera: large, circular-shaped opening formed when the top of a volcano collapses. (p. 174)

caldera: apertura grande circular que se crea cuando la cima de un volcán se colapsa. (p. 174)

GLOSSARY/GLOSARIO F ◆ 227

Glossary/Glosario

cementation: sedimentary rock-forming process in which sediment grains are held together by natural cements that are produced when water moves through rock and soil. (p. 51)

cinder cone volcano: steep-sided, loosely packed volcano formed when tephra falls to the ground. (p. 166)

cleavage: physical property of some minerals that causes them to break along smooth, flat surfaces. (p. 17)

coal: sedimentary rock formed from decayed plant material; the world's most abundant fossil fuel. (p. 67)

compaction: process that forms sedimentary rocks when layers of sediments are compressed by the weight of the layers above them. (p. 50)

composite volcano: volcano built by alternating explosive and quiet eruptions that produce layers of tephra and lava; found mostly where Earth's plates come together and one plate sinks below the other. (p. 167)

continental drift: Wegener's hypothesis that all continents were once connected in a single large landmass that broke apart about 200 million years ago and drifted slowly to their current positions. (p. 98)

convection current: current in Earth's mantle that transfers heat in Earth's interior and is the driving force for plate tectonics. (p. 111)

crater: steep-walled depression around a volcano's vent. (p. 158)

crystal: solid in which the atoms are arranged in an orderly, repeating pattern. (p. 9)

cementación: proceso de formación de la roca sedimentaria en el que las partículas de sedimento están unidas por cementos naturales producidos cuando el agua se mueve a través de la roca y el suelo. (p. 51)

volcán de cono de ceniza: volcán de laderas inclinadas, poco compactado, que se forma cuando la tefra cae al suelo. (p. 166)

exfoliación: propiedad física de algunos minerales que causa que se rompan junto a superficies planas y lisas. (p. 17)

carbón mineral: roca sedimentaria formada a partir de material vegetal descompuesto; es el combustible fósil más abundante en el mundo. (p. 67)

compactación: proceso que forma rocas sedimentarias cuando las capas de sedimento son comprimidas por el peso de las capas superiores. (p. 50)

volcán compuesto: volcán formado por explosiones alternantes y erupciones de baja intensidad que producen capas de tefra y lava; se encuentran principalmente donde se unen las placas continentales y una se sumerge bajo la otra. (p. 167)

deriva continental: hipótesis de Wegener respecto a que todos los continentes estuvieron alguna vez conectados en una gran masa terrestre única que se fraccionó cerca de 200 millones de años atrás y sus trozos se han movilizado lentamente a la deriva hasta sus posiciones actuales. (p. 98)

corriente de convección: corriente en el manto de la Tierra que transfiere calor en el interior de la Tierra y es la causa de la tectónica de placas. (p. 111)

cráter: depresión con paredes pronunciadas alrededor de la apertura volcánica. (p. 158)

cristal: sólido en el que los átomos están alineados en forma ordenada y repetitiva. (p. 9)

D

dike: igneous rock feature formed when magma is squeezed into a vertical crack that cuts across rock layers and hardens underground. (p. 173)

dique: característica de la roca ígnea formada cuando el magma es comprimido en una grieta vertical que cruza capas rocosas y se endurece en el subsuelo. (p. 173)

E

earthquake: vibrations produced when rocks break along a fault. (p. 127)

epicenter (EH pih sen tur): point on Earth's surface directly above an earthquake's focus. (p. 131)

terremoto: vibraciones producidas cuando las rocas se rompen a lo largo de una falla. (p. 127)

epicentro: punto de la superficie terrestre directamente encima del foco del terremoto. (p. 131)

Glossary/Glosario

extrusive/hydroelectric energy   **extrusivo/energía hidroeléctrica**

extrusive: describes fine-grained igneous rock that forms when magma cools quickly at or near Earth's surface. (p. 41)

extrusivo: describe rocas ígneas de grano fino que se forman cuando el magma se enfría rápidamente en o cerca de la superficie terrestre. (p. 41)

F

fault: surface along which rocks move when they pass their elastic limit and break. (p. 126)

focus: in an earthquake, the point below Earth's surface where energy is released in the form of seismic waves. (p. 130)

foliated: describes metamorphic rock, such as slate and gneiss, whose mineral grains line up in parallel layers. (p. 47)

fossil fuel: nonrenewable energy resource, such as oil and coal, formed over millions of years from the remains of dead plants and other organisms. (p. 66)

fracture: physical property of some minerals that causes them to break with uneven, rough, or jagged surfaces. (p. 17)

falla: área a lo largo de la cual las rocas se mueven cuando sobrepasan su límite elástico y se rompen. (p. 126)

foco: en un terremoto, el punto bajo la superficie terrestre donde se libera la energía en forma de ondas sísmicas. (p. 130)

foliado: describe rocas metamórficas, como pizarra y gneis, cuyas vetas minerales se alinean en capas paralelas. (p. 47)

combustible fósil: recurso energético no renovable, como el petróleo y el carbón mineral, formado durante millones de años a partir de restos de plantas y otros organismos muertos. (p. 66)

fractura: propiedad física de algunos minerales que causa que se rompan formando superficies irregulares, ásperas o dentadas. (p. 17)

G

gem: beautiful, rare, highly prized mineral that can be worn in jewelry. (p. 19)

geothermal energy: inexhaustible energy resource that uses hot magma or hot, dry rocks from below Earth's surface to generate electricity. (p. 78)

granitic: describes generally light-colored, silica-rich igneous rock that is less dense than basaltic rock. (p. 43)

gema: mineral hermoso, raro y altamente valorado que puede usarse como joya. (p. 19)

energía geotérmica: recurso energético inagotable que utiliza el magma caliente o las piedras secas calientes encontradas debajo de la superficie terrestre para producir electricidad. (p. 78)

granítica: roca ígnea rica en sílice, generalmente de color claro y menos densa que la rocas basáltica. (p. 43)

H

hardness: measure of how easily a mineral can be scratched. (p. 15)

hot spot: the result of an unusually hot area at the boundary between Earth's mantle and core that forms volcanoes when melted rock is forced upward and breaks through the crust. (p. 160)

hydroelectric energy: electricity produced by waterpower using large dams in a river. (p. 78)

dureza: medida de la facilidad con que un mineral puede ser rayado. (p. 15)

punto caliente: el resultado de un área extraordinariamente caliente en los límites entre el manto y el núcleo de la Tierra; forma volcanes cuando la roca fundida es empujada hacia arriba y se abre paso hacia la corteza. (p. 160)

energía hidroeléctrica: electricidad producida por la energía hidráulica generada mediante represas grandes construidas en los ríos. (p. 78)

GLOSSARY/GLOSARIO   **F** ◆ **229**

Glossary/Glosario

igneous rock/nonfoliated   **roca ígnea/no foliado**

I

igneous rock: rock formed when magma or lava cools and hardens. (p. 40)

intrusive: describes a type of igneous rock that generally contains large crystals and forms when magma cools slowly beneath Earth's surface. (p. 41)

roca ígnea: roca formada cuando se enfría y endurece el magma o la lava. (p. 40)

intrusivo: describe un tipo de roca ígnea que generalmente contiene cristales grandes y se forma cuando el magma se enfría lentamente por debajo de la superficie terrestre. (p. 41)

L

lava: molten rock that flows from volcanoes onto Earth's surface. (p. 40)

liquefaction: occurs when wet soil acts more like a liquid during an earthquake. (p. 141)

lithosphere (LIH thuh sfihr): rigid layer of Earth about 100 km thick, made of the crust and a part of the upper mantle. (p. 106)

luster: describes the way a mineral reflects light from its surface; can be metallic or nonmetallic. (p. 16)

lava: roca derretida que fluye de los volcanes hacia la superficie terrestre. (p. 40)

licuefacción: ocurre cuando el suelo húmedo se comporta como un líquido durante un terremoto. (p. 141)

litosfera: capa rígida de la Tierra de unos 100 kilómetros de profundidad, comprende la corteza y una parte del manto superior. (p. 106)

brillo: describe la forma en que un mineral refleja la luz desde su superficie; puede ser metálicos o no metálicos. (p. 16)

M

magma: hot, melted rock material beneath Earth's surface. (p. 11)

magnitude: measure of the energy released during an earthquake. (p. 140)

metamorphic rock: forms when heat, pressure, or fluids act on igneous, sedimentary, or other metamorphic rock to change its form or composition, or both. (p. 45)

mineral: naturally occurring inorganic solid that has a definite chemical composition and an orderly internal atomic structure. (p. 8)

mineral resources: resources from which metals are obtained. (p. 83)

magma: material rocoso fundido y caliente que se encuentra por debajo de la superficie terrestre. (p. 11)

magnitud: medida de la energía liberada durante un terremoto. (p. 140)

roca metamórfica: se forma cuando el calor, la presión o los fluidos actúan sobre una roca ígnea, sedimentaria u otra roca metamórfica para cambiar su forma, composición o ambas. (p. 45)

mineral: sólido inorgánico que se encuentra en la naturaleza, tiene una composición química definida y una estructura atómica ordenada. (p. 8)

recursos minerales: recursos a partir de los cuales pueden obtenerse metales. (p. 83)

N

natural gas: fossil fuel formed from marine organisms that is often found in tilted or folded rock layers and is used for heating and cooking. (p. 69)

nonfoliated: describes metamorphic rock, such as quartzite or marble, whose mineral grains grow and rearrange but generally do not form layers. (p. 48)

gas natural: combustible fósil formado a partir de organismos marinos y que a menudo se encuentra en capas rocosas inclinadas o plegadas; se usa para calefacción y para cocinar. (p. 69)

no foliado: describe rocas metamórficas, como la cuarcita o el mármol, cuyas vetas minerales se acumulan y reestructuran pero rara vez forman capas. (p. 48)

Glossary/Glosario

normal fault: break in rock caused by tension forces, where rock above the fault surface moves down relative to the rock below the fault surface. (p. 128)

nuclear energy: alternative energy source that is based on atomic fission. (p. 73)

falla normal: ruptura en la roca causada por fuerzas de tensión, donde la roca sobre la superficie de la falla se mueve hacia abajo con respecto a la roca debajo de la superficie de la falla. (p. 128)

energía nuclear: fuente de energía alternativa que se basa en la fisión atómica. (p. 73)

O

oil: liquid fossil fuel formed from marine organisms that is burned to obtain energy and used in the manufacture of plastics. (p. 69)

ore: deposit in which a mineral exists in large enough amounts to be mined at a profit. (pp. 23, 83)

petróleo: combustible fósil líquido formado a partir de organismos marinos; es quemado para obtener energía y se usa en la manufactura de plásticos. (p. 69)

mena: depósito en el que existen cantidades suficientes de un mineral para que la actividad minera sea rentable. (pp. 23, 83)

P

Pangaea (pan JEE uh): large, ancient landmass that was composed of all the continents joined together. (p. 98)

plate tectonics: theory that Earth's crust and upper mantle are broken into plates that float and move around on a plasticlike layer of the mantle. (p. 106)

plate: a large section of Earth's oceanic or continental crust and rigid upper mantle that moves around on the asthenosphere. (p. 106)

primary wave: seismic wave that moves rock particles back-and-forth in the same direction that the wave travels. (p. 131)

Pangea: masa terrestre extensa y antigua que estaba compuesta por todos los continentes unidos. (p. 98)

placa: gran sección de la corteza terrestre u oceánica y del manto rígido superior que se mueve sobre la astenosfera. (p. 106)

tectónica de placas: teoría respecto a que la corteza terrestre y el manto superior están fraccionados en placas que flotan y se mueven sobre una capa plástica del manto. (p. 106)

onda primaria: onda sísmica que mueve partículas rocosas en la misma dirección en que viaja la onda. (p. 131)

R

recycling: conservation method in which old materials are processed to make new ones. (p. 87)

reserve: amount of a fossil fuel that can be extracted from Earth at a profit using current technology. (p. 71)

reverse fault: break in rock caused by compressive forces, where rock above the fault surface moves upward relative to the rock below the fault surface. (p. 128)

rock: mixture of one or more minerals, rock fragments, volcanic glass, organic matter, or other natural materials; can be igneous, metamorphic, or sedimentary. (p. 36)

reciclaje: método de conservación en el que los materiales usados son procesados para producir otros nuevos. (p. 87)

reserva: depósito de un combustible fósil que puede extraerse de la Tierra y del cual, utilizando la tecnología actual, se obtienen utilidades. (p. 71)

falla inversa: ruptura en la roca causada por fuerzas de compresión, donde la roca sobre la superficie de la falla se mueve hacia arriba con respecto a la roca debajo de la superficie de la falla. (p. 128)

roca: mezcla de uno o más minerales, fragmentos de roca, obsidiana, materia orgánica u otros materiales naturales; puede ser ígnea, metamórfica o sedimentaria. (p. 36)

Glossary/Glosario

rock cycle: model that describes how rocks slowly change from one form to another through time. (p. 37)

ciclo de la roca: modelo que describe cómo cambian lentamente las rocas de una forma a otra a través del tiempo. (p. 37)

S

seafloor spreading: Hess's theory that new seafloor is formed when magma is forced upward toward the surface at a mid-ocean ridge. (p. 103)

expansión del suelo oceánico: teoría de Hess respecto a que se forma un nuevo suelo oceánico cuando el magma es empujado hacia la superficie a través de un surco en la mitad del océano. (p. 103)

secondary wave: seismic wave that moves rock particles at right angles to the direction of the wave. (p. 131)

onda secundaria: onda sísmica que mueve partículas rocosas en ángulos rectos respecto a la dirección de la onda. (p. 131)

sedimentary rock: forms when sediments are compacted and cemented together or when minerals form from solutions. (p. 49)

roca sedimentaria: se forma cuando los sedimentos son compactados y cementados o cuando se forman minerales a partir de soluciones. (p. 49)

sediments: loose materials, such as rock fragments, mineral grains, and the remains of once-living plants and animals, that have been moved by wind, water, ice, or gravity. (p. 49)

sedimentos: materiales sueltos, como fragmentos de roca, granos minerales y restos de animales y plantas, que han sido arrastrados por el viento, el agua, el hielo o la gravedad. (p. 49)

seismic (SIZE mihk) wave: wave generated by an earthquake. (p. 130)

onda sísmica: onda generada por un terremoto. (p. 130)

seismograph: instrument used to register earthquake waves and record the time that each arrived. (p. 133)

sismógrafo: instrumento utilizado para registrar las ondas sísmicas y la hora a la que llega cada una. (p. 133)

shield volcano: broad, gently sloping volcano formed by quiet eruptions of basaltic lava. (p. 166)

volcán de escudo: volcán levemente inclinado y de gran extensión, formado por erupciones de baja intensidad de lava basáltica. (p. 166)

silicate: mineral that contains silicon and oxygen and usually one or more other elements. (p. 12)

silicato: mineral que contiene sílice y oxígeno y generalmente uno o varios elementos distintos. (p. 12)

sill: igneous rock feature formed when magma is squeezed into a horizontal crack between layers of rock and hardens underground. (p. 173)

alféizar: roca ígnea característica formada cuando el magma es comprimido en una grieta horizontal entre capas de roca y se endurece en el subsuelo. (p. 173)

solar energy: energy from the Sun that is clean, inexhaustible, and can be transformed into electricity by solar cells. (p. 76)

energía solar: energía del sol, la cual es limpia e inagotable y puede transformarse en electricidad a través de celdas solares. (p. 76)

specific gravity: ratio of a mineral's weight compared with the weight of an equal volume of water. (p. 16)

gravedad específica: cociente del peso de un mineral comparado con el peso de un volumen igual de agua. (p. 16)

streak: color of a mineral when it is in powdered form. (p. 17)

veta: color de un mineral en forma de polvo. (p. 17)

strike-slip fault: break in rock caused by shear forces, where rocks move past each other without much vertical movement. (p. 129)

falla deslizante: ruptura en la roca causada por fuerzas opuestas, donde las rocas se mueven una tras otra sin mucho movimiento vertical. (p. 129)

surface wave: seismic wave that moves rock particles up-and-down in a backward rolling motion and side-to-side in a swaying motion. (p. 131)

onda de superficie: onda sísmica que mueve partículas rocosas en forma ascendente y descendente en un movimiento circular en retroceso y de un lado a otro en un movimiento oscilante. (p. 131)

Glossary/Glosario

tephra/wind farm — **tefra/granja de energía eólica**

tephra (TEFF ruh): bits of rock or solidified lava dropped from the air during an explosive volcanic eruption; ranges in size from volcanic ash to volcanic bombs and blocks. (p. 166)

tsunami (soo NAH mee): seismic sea wave that begins over an earthquake focus and can be highly destructive when it crashes on shore. (p. 142)

tefra: trozos de roca o lava solidificada que caen del aire durante una erupción volcánica explosiva; su tamaño oscila desde la ceniza volcánica hasta las bombas o bloques volcánicos. (p. 166)

maremoto: onda sísmica marina que comienza sobre el foco del terremoto y que puede ser altamente destructiva cuando se estrella en la costa. (p. 142)

vent: opening where magma is forced up and flows out onto Earth's surface as lava, forming a volcano. (p. 158)

volcanic neck: solid igneous core of a volcano left behind after the softer cone has been eroded. (p. 173)

volcano: opening in Earth's surface that erupts sulfurous gases, ash, and lava; can form at Earth's plate boundaries, where plates move apart or together, and at hot spots. (p. 156)

chimenea: apertura donde el magma es empujado hacia arriba y fluye sobre la superficie terrestre como lava, formando un volcán. (p. 158)

cuello volcánico: núcleo ígneo sólido de un volcán que queda después de que el cono más blando ha sido erosionado. (p. 173)

volcán: apertura en la superficie terrestre que arroja gases sulfurosos, ceniza y lava; puede formarse en los límites de las placas continentales, donde las placas se separan o encuentran y en los puntos calientes. (p. 156)

wind farm: area where many windmills use wind to generate electricity. (p. 77)

granja de energía eólica: área en donde muchos molinos usan el viento para generar electricidad. (p. 77)

Index

Abrasives   Core(s)

Italic numbers = illustration/photo   **Bold numbers** = **vocabulary term**
lab = a page on which the entry is used in a lab
act = a page on which the entry is used in an activity

A

Abrasives, 85
Activities, Applying Math, 54, 143, 172; Applying Science, 16, 86, 108; Integrate Career, 77, 118; Integrate Chemistry, 23, 43, 52, 84, 103; Integrate Environment, 158; Integrate Health, 166; Integrate Life Science, 67; Integrate Physics, 11, 39, 77, 103, 114, 128, 131, 141; Science Online, 22, 42, 46, 79, 99, 108, 133, 142, 163, 173; Standardized Test Practice, 32–33, 62–63, 94–95, 122–123, 152–153, 182–183
Aggregate, 86
Alaska, earthquake in, 148; volcanoes in, *165*, 168
Alcohol, energy from, 80, *80*
Alluvial fan, *38*
Almandine, 21, *21*
Alternative resources, biomass energy, *79*, 79 act, 79–81, *80*, *81*; geothermal energy, 78, *79*; hydroelectric power, 78, *78*; nuclear energy, 73, 73–75, *74*, *75*; solar energy, 76, 76–77, 82 lab; wind energy, 77, *77*
Aluminum, 23, *23*, 83
Amethyst, 19, 21, *21*
Andesite, 42
Andesitic magma, 165
Andesitic rock, *42*, 43
Animal(s), and earthquakes, 148, *148*
Apatite, 15
Applying Math, Chapter Review, 31, 61, 93, 121, 151, 181; Classifying Igneous Rocks, 172; Coal Formation, 54; Earthquake Energy, 143; Section Review, 25, 43, 55, 75, 104, 137, 175

Applying Science, How well do the continents fit together?, 108; Mineral Identification, 16; Why should you recycle?, 86
Applying Skills, 12, 18, 39, 48, 81, 87, 101, 115, 129, 145, 161, 169
Ash, volcanic, 165, 178, *178*
Asthenosphere, 106, 135, 137
Ayers rock (Australia), 58, *58*
Azurite, *14*

B

Barite, *10*
Basalt, *42*, *43*, 44 lab
Basaltic magma, 163, 166
Basaltic rock, *42*, **43**
Batholiths, 172
Bauxite, 23, *23*, 83
Beryl, 20, *20*
Biomass energy, *79*, 79 act, 79–81, *80*, *81*
Bituminous coal, 68, *68*
Breccia, *51*, *52*
Building(s), quake-resistant, 144, *144*, 144 lab
Building materials, 86

C

Calcite, 12, 15, 16, 18, 36, 48, 51, 52, 53
Calcium carbonate, 53
Caldera, 174, *174*, 176–177 lab
Cave(s), formation of, 52
Cell(s), solar, 76, *76*
Cementation, 51, *51*
Chalk, *38*, 53
Chemical sedimentary rocks, 52–53, **53**
Cinder cone volcano, 166–167, *167*, 175
Classification, of igneous rock, *42*,

42–43; of igneous rocks, 172; of metamorphic rocks, 47, 47–48, *48*; of sedimentary rocks, 50–57, *51*, 56–57 lab; of sediments, 50 lab
Clay, *51*
Cleavage, 17, *17*
Climate, as evidence of continental drift, 100
Coal, 54, 54 act, 55, **67,** *67*; bituminous, 68, *68*; conservation of, 73; formation of, 67, 68, *68*; lignite, 68, *68*; mining, 70, *70*; uses of, 66
Color, of minerals, 16
Columbia Plateau, 166
Communicating Your Data, 13, 27, 44, 57, 89, 111, 117, 147, 170, 177
Compaction, 50, *50*
Composite volcano, 167, *167*
Compression forces, 113, 128, *128*
Concept Mapping, 31
Concrete, 52, *52*, 86
Cone(s), cinder cone volcanoes, **166**–167, *167*, 175; modeling, 166 lab
Conglomerate, 51, *51*, *52*, 52
Conservation, and energy-efficient housing, 88–89 lab; of fossil fuels, 73, 73 act; and recycling, 86, 86 act, 87
Continent(s), fitting together, 97 lab, 108, 108 act
Continental drift, 98–101, 99 act; course of, 101, *101*; evidence for, 97 lab, 98, 99, 100, 100 lab
Convection current, 111, *111*, 111 act
Convergent plate boundaries, 108, 109, *109*, 110, 113, 160, *160*
Copper, 16 act
Coquina, 53
Core(s), inner, 135, *135*; outer, 135, *135*

Index

C (cont.)

Corundum, 15, 21, *21*, 85
Crater(s), 158, *158*
Crater Lake (Oregon), 174, *174, 175*
Crust, 135, *135;* of Earth, 12, *12*
Crystal, 9, **9**–11, *10, 11,* 13 *lab*
Crystalline, 8
Crystal systems, 9 *lab, 10*
Curie point, 103
Cycles, rock, *37,* **37**–39, *38,* 55

Dam, 78, *78*
Data Source, 88, 116
Design Your Own, How do calderas form?, 176–177; Mineral Identification, 26–27
Detrital sedimentary rocks, 50–52, *51*
Diabetes, 28
Diamond, 6, 12, 15, *15,* 22, *22*
Dike, 173, *173*
Diorite, *42*
Divergent plate boundaries, 107, 109, *109,* 159, *159*
Dolomite, 12
Drift mining, 70
Drilling, 71, *71,* 90, *90*

Earth, crust of, 135, *135;* inner core of, 135, *135;* lithosphere of, 106, *106, 107;* magnetic field of, 97–104, *104;* magnetism of, 141; mantle of, 111, *111,* 135, *135,* 137, *137;* minerals in crust of, 12, *12;* outer core of, 135, *135;* seismic wave studies of, 135 *lab, 136,* 136–137; structure of, *135,* 135–137, *136*
Earthquakes, 108 *act,* 124–148, **127,** 133 *lab;* causes of, 125 *lab, 126,* 126–127, *127,* 146–147 *lab;* damage caused by, 124, 139, *139,* 140, *140,* 141, *141,* 148, *148;* energy of, 143 *act;* epicenter of, *127, 132,* 133–134, *134,* 138 *lab;* and faults. *See* Faults; features of, 130–137;

focus of, 130, *132,* 146–147 *lab;* hazard map of, 143, *143;* intensity of, 141, 148; magnitude of, 140; safety and, 143–145, *144,* 144 *lab, 145;* and seismic waves, *130,* 130–132, *131, 132, 133,* 133–134, *134,* 135 *lab, 136, 136*
East African Rift Valley, 96
East Pacific Rise, 112
Elastic deformation, 126
Elastic limit, 126
Electricity, sources of. *See* Energy sources
Emerald, 20, *20*
Energy, biomass, *79,* 79 *act,* **79**–81, *80, 81;* from fusion, 75, *75;* geothermal, **78,** *79;* hydroelectric, **78,** *78;* nuclear, *73,* **73**–75, *74, 75;* solar, *76,* **76**–77, 82 *lab;* from water, 78, *78;* wind, 77, *77*
Energy sources, fossil fuels, 66–73; inexhaustible, *76,* 76–79, *77, 78, 79;* nonrenewable, 66–75; nuclear, *73,* **73**–75, *74, 75;* percentage used in U.S., *69;* renewable, 76–82
Environment, effects of volcanic eruptions on, *156,* 157, *157*
Environmental Protection Agency, 74
Epicenter, *127, 132,* **133**–134, *134,* 138 *lab*
Erosion, of rocks, 50
Eruptions, effects of, *156,* 157, *157;* factors in, 162–163; quiet, 163; violent, 162–163, 165, *165,* 168–169, *169*
Etna, Mount (Italy), 154, *154,* 168
Evaporites, 11, *11*
Extrusive rock, 41, *41, 42*

Fault(s), 110, *110,* **126,** *132;* causes of, 125 *lab, 127, 127;* formation of, 125 *lab,* 126, *126;* normal, 112; strike-slip, 114, *114;* types of, *128,* 128–129, *129*
Fault-block mountains, 112, *112*

Feldspar, 12, 15, 36, *36,* 47, 52
Fertilizer, 85
Fission, nuclear, **73**–74, *74*
Fissure, 166
Fluorite, *10,* 15, *16*
Focus, 130, *132,* 146–147 *lab*
Foldables, 7, 35, 65, 97, 125, 155
Foliated rocks, 47, *47*
Fool's gold (pyrite), 14, *14,* 17
Force(s), compression, 113, **128,** *128;* direction of, 114
Fossil fuels, 66–73; coal, 66, 67, *67, 68, 68,* 70, *70,* 73; conserving, 73, 73 *act;* methane hydrates, 71, *72;* natural gas, 66, 69, *69,* 71, *71,* 73; as nonrenewable resources, 66; oil, 65 *lab,* 66, 69, *69,* 71, *71,* 73, 73 *act,* 90, *90;* removing from ground, *70,* 70–71, *71,* 90, *90;* reserves of, 71; uses of, 66
Fossil record, as evidence of continental drift, 99, *99, 100,* 100 *lab*
Fossil-rich limestone, 53
Fracture, 17
Fuel, fossil. *See* Fossil fuels; synthetic, 67
Fusion, nuclear, 75, *75*

Gabbro, *42,* 44 *lab*
Galina, 16 *act*
Galvanization, 24
Gangue, 84
Garnet, 21, *21,* 85
Gas(es), natural, 66, 69, *69,* 71, *71,* 73; trapped in volcanoes, 162
Gasohol, 80, *80*
Gems, *19,* **19**–22, *20, 21, 22,* 22 *act*
Geothermal energy, 78, *79*
Glaciers, as evidence of continental drift, 100
Glomar Challenger **(research ship),** 103
Gneiss, *38, 45,* 46, 47, 52
Gold, 16 *act;* identifying, 14, *14,* 17
Granite, 36, *36, 42, 43,* 44 *lab,* 47, 52, 86
Granitic magma, 165, *165*

Index

Granitic rock

Granitic rock, *42*, 43
Graphite, 8, *16*, 17
Gravel, 51
Great Rift Valley, 107, 112
Gypsum, 10, 11, 15, 52, 86

Halite, 8, 11, *11*, 17, 53, *53*, 85, *85*. *See also* Salt(s)
Hardness, 15
Hawaiian Islands, volcanoes in, 156, 160, 161, *161*, 163, *166*, 168
Hekla volcano (Iceland), 156, *156*
Hematite, 16 *act*, 17, 23, 48, 51, 83
Herculaneum (Italy), 178, *178*
Hess, Harry, 103
Himalaya, 113, *113*
Hodgkin, Dorothy Crowfoot, 28, *28*
Hornblende, 36, 48
Hot spots, 160–161, *161*
Housing, energy-efficient, 88–89 *lab*
Hutton, James, 39
Hydrocarbons, 66, 67
Hydroelectric energy, 78, *78*

Iceland, volcanoes in, 156, *156*, 159, 160, 168
Igneous rocks, 37, *37*, 38, **40–44**, 171–175, *173*, *174*; classifying, 42, 42–43, 172; formation of, 40, 40–42, 41, 44 *lab*; intrusive features of, *171*, 171–173, *173*, 173 *act*
Ilmenite, 24, *24*
Indonesia, volcanoes in, 165, 168, 169, *169*
Industrial minerals, 85, *85*
Inexhaustible resources, 76, 76–79, *77*, *78*, *79*
Inner core, 135, *135*
Insulation, 85, 85 *lab*
Insulin, 28
Integrate Astronomy, volcano, 158
Integrate Career, 77; volcanologist, 118
Integrate Chemistry, 23;
cave formation, 52; Curie point, 103; melting rock, 43; refining ore, 84
Integrate Environment, environmental impacts, 157
Integrate Health, volcanic ash, 165
Integrate Life Science, coal formation, 67
Integrate Physics, crystal formation, 11; direction of forces, 114; magnetic clues, 103; magnetism, 141; matter and the rock cycle, 39; sound waves, 131; types of faults, 128; wind energy, 77
Intrusive igneous rock features, *171*, 171–173, *173*, 173 *act*
Intrusive rock, 41, *41*, *42*
Io (moon of Jupiter), 158
Iron, 19; as nonrenewable resource, 83, 84, *84*; in ore, 23
Italy, volcanoes in, 154, *154*, 168, 178, *178*

Journal, 6, 34, 64, 96, 124, 154
Jupiter, moons of, 158

Kauai volcano (Hawaiian Islands), 161
Kilauea volcano (Hawaiian Islands), 156, 168
Krakatau volcano (Indonesia), 165, 168, 169, *169*

Lab(s), Crystal Formation, 13; Design Your Own, 26–27, 176–177; Earthquake Depths, 146–147; Epicenter Location, 138; Identifying Types of Volcanoes, 170; Igneous Rock Clues, 44; Launch Labs, 7, 35, 65, 97, 125, 155; MiniLabs, 18, 50, 85, 111, 135, 166; Model and Invent, 88–89; Predicting Tectonic Activity, 116–117;
Seafloor Spreading Rates, 105; Sedimentary Rocks, 56–57; Soaking Up Solar Energy, 82; Try at Home MiniLabs, 9, 37, 73, 100, 144, 160; Use the Internet, 116–117
Labradorite, 11
Lasers, 75
Launch Labs, Distinguishing Rocks from Minerals, 7; Finding Energy Resources, 65; Map a Volcano, 155; Observe and Describe Rocks, 35; Were the continents connected?, 97; Why do earthquakes occur?, 125
Lava, 40–41, *41*, 42 *act*, *158*, 163, *164*; layers of, 167
Lignite, 68, *68*
Limestone, 48, 52, 53, 86
Liquefaction, 141, *141*
Lithosphere, 106, *106*, 107, 135
Lodestone, 18, *18*
Luster, 16, *16*

Magma, 11, *11*, 37, 40, 40–41, *41*, 42, 103, 158, *158*, 160, *160*; andesitic, 165; basaltic, 163, 166; composition of, 163–167; granitic, 165, *165*; movement of, 163 *act*
Magnetic field(s), of Earth, 103–104, *104*; and seafloor spreading, 103–104, *104*
Magnetic properties, 18, *18*
Magnetic time scale, 104
Magnetism, of Earth, 141
Magnetite, 16 *act*, 18, *18*, 23, 104
Magnetometer, 104
Magnitude, 140
Mantle, 135, *135*, 137, *137*; of Earth, 111, *111*
Map(s), hazard map of earthquakes, 143, *143*; topographic, 155 *lab*; of volcanoes, 155 *lab*
Marble, 48, *48*
Matter, and rock cycle, 39
Mauna Loa volcano (Hawaiian Islands), 166

Mazama, Mount (Oregon), 174
Measurement, of earthquake magnitude, 140
Medicine, insulin, 28
Metallic mineral resources, 83, 83–84, 84
Metamorphic rocks, 37, 37, 38, 45–48; classifying, 47, 47–48, 48; formation of, 45, 45–46, 46
Methane, 67
Methane hydrates, 71, 72
Mexico, volcanoes in, 167, 167, 168
Mica, 17, 36, 36, 47
Mid-Atlantic Ridge, 107, 108, 112, 159
Mid-ocean ridges, 102, 102, 107, 108, 112
Mineral(s), 6–27, 8; appearance of, 14, 14; characteristics of, 8; cleavage of, 17, 17; color of, 16; distinguishing rocks from, 7 lab; in Earth's crust, 12, 12; fracture of, 17; gems, 19, 19–22, 20, 21, 22, 22 act; hardness of, 15; identifying, 14, 14–18, 16 act, 18, 26–27 lab; industrial, 85, 85; luster of, 16, 16; magnetic properties of, 18, 18; physical properties of, 14, 14–18, 16, 17, 18, 26–27 lab; rock-forming, 12; streak test of, 17, 17; structure of, 9, 9–11, 10, 11, 13 lab; unique properties of, 18; useful elements in, 23, 23–25, 24, 25; uses of, 8, 8, 19–25; vein, 24, 24
Mineral grains, 41, 44 lab, 45
Mineral resources, 83–87; metallic, 83, 83–84, 84; nonmetallic, 84–86; recycling, 86, 86 act, 87
MiniLabs, Classifying Sediments, 50; Interpreting Seismic Wave Data, 135; Modeling Convection Currents, 111; Modeling Volcanic Cones, 166; Observing Mineral Properties, 18; Observing the Effects of Insulation, 85
Mining, 70, 70
Model and Invent, Home Sweet Home, 88–89
Modified Mercalli intensity scale, 141

Mohorovicic, Andrija, 137
Mohorovicic discontinuity (Moho), 137, 137
Mohs, Friedrich, 15
Mohs scale of hardness, 15
Montserrat volcano, 157, 157, 160, 162, 165, 167, 168
Moon(s), of Jupiter, 158
Mountains, as evidence of continental drift, 100; fault-block, 112, 112; formation of, 113, 113
Mount St. Helens eruption (Washington state), 162, 162–163, 168

National Geographic Visualizing, Crystal Systems, 10; Lava, 164; Methane Hydrates, 72; Plate Boundaries, 109; The Rock Cycle, 38; Seismic Waves, 132
Natural gas, 69; conservation of, 73; drilling for, 71, 71; formation of, 69, 69; uses of, 66
Nonfoliated rocks, 48, 48
Nonmetallic mineral resources, 84–86
Nonrenewable resources, 66–75
Normal fault, 112, 128, 128
Northridge earthquake, 139, 139, 141
Nuclear energy, 73, 73–75, 74, 75
Nuclear fission, 73–74, 74
Nuclear fusion, 75, 75
Nuclear reactors, 74, 74

Obsidian, 42, 44 lab
Ocean floor, mapping, 102; spreading of, 102, 103–104, 105 lab
Oil (petroleum), 69; conservation of, 73, 73 act; discovery of, 90, 90; drilling for, 71, 71, 90, 90; formation of, 69, 69; and rock, 65 lab; uses of, 66
Olivine, 21, 21

Oops! Accidents in Science, Black Gold, 90; Buried in Ash, 178
Opal, 8
Open-pit mining, 70, 70
Ore, 23, 23, 83–84, 84
Organic sedimentary rocks, 53–54, 55
Ortelius, Abraham, 98
Outer core, 135, 135
Oxygen, 12

Pahoehoe lava, 163, 164
Pangaea, 98, 98, 99
Paricutín volcano (Mexico), 167, 167, 168
Peat, 68, 68
Percentages, 172 act
Peridot, 21, 21
Petroleum. See Oil (petroleum)
Phyllite, 46
Physical properties, appearance, 14, 14; cleavage, 17, 17; color, 16; fracture, 17; hardness, 15; luster, 16, 16; of minerals, 14, 14–18, 16, 17, 18, 26–27 lab; streak, 16, 17, 17
Pillow lava, 164
Pinatubo volcano (Philippines), 162, 168
Plant(s), as evidence of continental drift, 99, 100, 100; and volcanoes, 157
Plate(s), 106, 107, 127, 127, 129, 129, 146–147 lab; collision of, 109, 110; composition of, 106, 106
Plate boundaries, 107, 107; convergent, 108, 109, 109, 110, 113, 160, 160; divergent, 107, 109, 109, 159, 159; transform, 110, 110
Plate tectonics, 96, 106–117; causes of, 111, 111; features caused by, 112, 112–114, 113, 114; predicting activity, 116–117 lab; testing for, 114, 114–115
Pompeii (Italy), volcanic eruption in, 178
Pressure, and metamorphic rocks, 46

Index

Primary wave, 131, *131, 132,* 134, *134, 136, 136*
Properties, of gems, 19; magnetic, 18, *18;* physical. *See* Physical properties
Pumice, *42,* 44 *lab*
Pyrite (fool's gold), 14, *14,* 17

Q

Quartz, 9, *9,* 12, 15, 16, 17, 21, *21,* 22, 36, *36,* 47, 48, 51, 52
Quartzite, 48

R

Rainier, Mount (Washington state), *38,* 167
Ratios, 143 *act*
Reading Check, 9, 15, 17, 23, 24, 37, 39, 41, 42, 46, 47, 50, 52, 67, 71, 74, 77, 80, 83, 86, 98, 99, 102, 103, 107, 112, 113, 127, 129, 131, 135, 136, 144, 158, 165
Real-World Questions, 13, 26, 44, 56, 82, 88, 105, 116, 138, 146, 170, 176
Recycling, 86, 86 *act,* 87
Refining ore, 84, *84*
Renewable resources, 76–82
Reserve, 71
Reservoir rock, 69
Resources, conservation of, 73, 73 *act;* inexhaustible, 76, 76–79, *77, 78, 79;* mineral, **83**–87; nonrenewable, 66–75; renewable, 76–82
Reverse fault, 128, *128*
Rhodonite, 10
Rhyolite, *42,* 44 *lab*
Richter scale, 140, 143 *act*
Rift valleys, 96, 107, *109,* 112
Ring of fire, 159
Rock(s), 34–58, **36;** andesitic, *42, 43;* basaltic, *42,* **43;** as building material, 86; cementation of, 51, *51;* common, 36, *36;* compaction of, 50, *50;* distinguishing minerals from, 7 *lab;* erosion of, 50; as evidence of continental drift, 100; extrusive, **41,** *41, 42;* foliated, **47,** *47;* granitic, *42,* **43;** igneous, 37, *37, 38, 40,* **40**–44, *41, 42,* 44 *lab, 171,* 171–175, *173,* 173 *act;* intrusive, **41,** *41, 42;* melting, 43; metamorphic, 37, *37, 38, 45,* **45**–48, *46, 47, 48;* modeling, 37 *lab;* nonfoliated, **48,** *48;* observing and describing, 35 *lab;* and oil, 65 *lab;* reservoir, 69; sedimentary, 37, *37,* **39**–57, 56–57 *lab;* stacked, 49, *49;* structure of, 11, *11;* weathering of, 50
Rock cycle, 37, *37*–39, *38,* 55
Rock-forming minerals, 12
Rock gypsum, 52
Rock salt, 53, *53*
Rubies, 20, *20,* 22
Rushmore, Mount, 36
Rutile, 24, *24*

S

Safety, earthquake, 143–145, *144,* 144 *lab, 145*
Salt(s), crystal structure of, 9, 9 *lab;* from halite, 85, *85;* road, *85;* rock, 53, *53;* uses of, 8
San Andreas Fault, 110, *110,* 114, 129, *129*
Sand, *51*
Sandpaper, 85
Sandstone, 48, *51,* 52, 85, 86
San Francisco earthquake, 140, *141,* 143
Sapphire, 21, *21*
Satellite Laser Ranging System, 114, *114*
Schist, 46
Science and History, Dr. Dorothy Crowfoot Hodgkin, 28, *28*
Science and Language Arts, Listening In, 118
Science and Society, Australia's Controversial Rock Star, 58, *58*
Science Online, biomass energy, 79; continental drift, 99; earthquake data, 133; earthquakes and volcanoes, 108; gemstone data, 22; igneous rock features, 173; Kilauea volcano, 163; rock formation, 42; shale metamorphism, 46; tsunamis, 142
Science Stats, 148
Scientific Methods, 13, 26–27, 44, 56–57, 82, 88–89, 105, 116–117, 138, 146–147, 170, 176–177; Analyze Your Data, 27, 89, 117, 146, 177; Conclude and Apply, 13, 27, 44, 57, 82, 89, 105, 117, 138, 147, 170, 177; Follow Your Plan, 177; Form a Hypothesis, 26, 116, 177; Make a Plan, 177; Make the Model, 89; Plan the Model, 89; Test Your Hypothesis, 117; Test Your Model, 89
Scoria, *42,* 167
Seafloor spreading, *102,* **103**–104, 105 *lab*
Secondary wave, 131, *132,* 134, *134, 136, 136*
Sediment(s), 49; classifying, 50 *lab;* size and shape of, 51, *51*
Sedimentary rocks, 37, *37,* **39**–57; chemical, 52–53, *53;* classifying, 50–57, *51,* 56–57 *lab;* detrital, 50–52, *51;* formation of, 49, *49;* materials found in, 52, *52;* organic, 53–54, *55*
Seismic sea waves (tsunamis), 142, *142,* 148
Seismic waves, 130, 130–132, *131, 132, 133,* 133–134, *134,* 135, *136, 136*
Seismograph, *133,* **133**–134, *134,* 140
Serpentine, 48
Shadow zones, 136, *136*
Shale, 38, 46 *act, 51,* 52
Shield volcano, 166, *166*
Ship Rock (New Mexico), 173
Siccar Point (Scotland), 39, *39*
Silica, 85
Silica-rich magma, 165, *165*
Silicates, 12
Silicon, 12
Sill, 173, *173*
Silt, *51*
Siltstone, *51,* 52
Silver, 16 *act*
Slate, 46, 47, *47*
Slope mines, 70
Smelting, 23, 84, *84*
Sodium chloride, 8. *See also* Salt(s)
Solar cells, 76, *76*
Solar energy, 76, **76**–77, 82 *lab*

Index

Solution

Solution, crystals from, 11, *11*
Soufrière Hills volcano (Montserrat), 157, *157*, 160, 162, 165, 167, 168
Sound waves, 131
Sphalerite, 24, *24*
Spinel, 20, *20*
Stacked rock, 49, *49*
Standardized Test Practice, 32–33, 62–63, 94–95, 122–123, 152–153, 182–183
Streak, 16, 17, *17*
Strike-slip fault, 114, *114,* **129,** *129*
Study Guide, 29, 59, 91, 119, 149, 179
Subduction zones, 108, 110
Submarine, nuclear, 73, *73*
Sun, energy from, *76,* 76–77, 82 *lab*
Surface wave, 131, *131, 132*
Surtsey volcano (Iceland), 159, 168
Sylvite, 85
Synthetic fuels, 67

Talc, 15
Tanzanite, 20, *20*
Technology, dams, 78, *78;* gasohol, 80, *80; Glomar Challenger* (research ship), 103; hydroelectric power, 78, *78;* lasers, *75;* magnetometer, 104; nuclear fusion, 75, *75;* nuclear reactors, 74, *74;* nuclear submarine, 73, *73;* refining ore, 84, *84;* Satellite Laser Ranging System, 114, *114;* seismic wave studies of Earth, 135 *lab,* 136, 136–137; seismographs, *133,* 133–134, *134,* 140; smelting, 84, *84;* solar cells, 76, *76;* testing for plate tectonics, 114–115, *115;* wheelchairs, 25, *25;* windmills, 77, *77;* X-ray crystallography, 28, *28*

Tectonic plate(s). *See* Plate(s); Plate tectonics
Temperature, and metamorphic rocks, 46, *46*
Tephra, 166, *167*
Thickness, calculating, 54 *act*
TIME, Science and History, 28, *28;* Science and Society, 58
Titanium, 24, 24–25, *25*
Topaz, 15, 20, *20*
Topographic maps, 155 *lab*
Transform plate boundaries, 110, *110*
Tripp, Charles, 90
Trona, 85
Try at Home MiniLabs, Inferring Salt's Crystal System, 9; Interpreting Fossil Data, 100; Modeling Magma Movement, 160; Modeling Rock, 37; Modeling Seismic-Safe Structures, 144; Practicing Energy Conservation, 73
Tsunami, 142, *142,* 142 *act,* 148
Tube worms, 103

Uluru (Australia), 58, *58*
Underground mining, 70, *70*
Uranium, 74
Use the Internet, Predicting Tectonic Activity, 116–117

Vein mineral deposits, 24, *24*
Vent, 158
Vesuvius volcano (Italy), 168, 178, *178*
Volcanic glass, 42
Volcanic neck, 173, *173*
Volcano(es), 154–180, **166,** *166;* ash from, 165, 178, *178;* calderas formed by, 174, *174,* 176–177 *lab;* eruptions of, *156,*

Zoisite

157, *157,* 162–163, 165, 168–169, *169;* formation of, 158, *158;* and formation of igneous rock, *40,* 40–41, *41;* forms of, *166,* 166–169, *167,* 170 *lab;* gases trapped in, 162; igneous rock features of, *171,* 171–175, *173,* 173 *act, 174;* location of, *159,* 159–161; mapping, 155 *lab;* and plate tectonics, 108 *act,* 110, 113
Volcanologist, 118

Waste(s), energy from, 81, *81*
Water, energy from, 78, *78*
Water vapor, and volcanic eruption, 163
Wave(s), primary, **131,** *131,* 132, 134, *134,* 136, *136;* secondary, **131,** *132,* 134, *134,* 136, *136;* seismic, *130,* **130**–132, *131, 132, 133,* 133–134, *134,* 135 *lab,* 136, *136;* sound, 131; surface, **131,** *131,* 132; tsunami, 142, *142,* 142 *act,* 148
Weathering, of rocks, 50
Wegener, Alfred, 98, 99, 100, 101
Wheelchairs, 25, *25*
Wind energy, 77, *77*
Wind farms, 77, *77*
Windmills, 77, *77*
Wood, energy from, 80, *80*

X-ray crystallography, 28, *28*

Zinc, 24
Zircon, *10*
Zoisite, 20, *20*

Credits

Magnification Key: Magnifications listed are the magnifications at which images were originally photographed.
LM–Light Microscope
SEM–Scanning Electron Microscope
TEM–Transmission Electron Microscope

Acknowledgments: Glencoe would like to acknowledge the artists and agencies who participated in illustrating this program: Absolute Science Illustration; Andrew Evansen; Argosy; Articulate Graphics; Craig Attebery, represented by Frank & Jeff Lavaty; CHK America; John Edwards and Associates; Gagliano Graphics; Pedro Julio Gonzalez, represented by Melissa Turk & The Artist Network; Robert Hynes, represented by Mendola Ltd.; Morgan Cain & Associates; JTH Illustration; Laurie O'Keefe; Matthew Pippin, represented by Beranbaum Artist's Representative; Precision Graphics; Publisher's Art; Rolin Graphics, Inc.; Wendy Smith, represented by Melissa Turk & The Artist Network; Kevin Torline, represented by Berendsen and Associates, Inc.; WILDlife ART; Phil Wilson, represented by Cliff Knecht Artist Representative; Zoo Botanica.

Photo Credits

Cover PhotoDisc; **i ii** PhotoDisc; **iv** (bkgd)John Evans, (inset)PhotoDisc; **v** (t)PhotoDisc, (b)John Evans; **vi** (l)John Evans, (r)Geoff Butler; **vii** (l)John Evans, (r)PhotoDisc; **viii** PhotoDisc; **ix** Aaron Haupt Photography; **x** Inga Spence/Visuals Unlimited; **xi** Soames Summerhays/Photo Researchers; **xii** (tl)Albert J. Copley/Visuals Unlimited, (tc bcr)Mark A. Schneider/Visuals Unlimited, (tr bcl)Doug Martin, (bl)Visuals Unlimited, (br)Jose Manuel Sanchis Calvete/CORBIS; **1** David J. Cross/Peter Arnold, Inc.; **2** (t)AP/Wide World Photos/Jack Smith, (b)AP/Wide World Photos/Gary Stewart; **3** (t)Francois Gohier/Photo Researchers, (b)David Muench/CORBIS; **5** (t)Chlaus Lotscher/Stock Boston, (b)USGS; **6–7** SuperStock; **8** Matt Meadows; **9** (inset)John R. Foster/Photo Researchers, (l)Mark A. Schneider/Visuals Unlimited; **10** (tr)Mark A. Schneider/Visuals Unlimited, (cl)A.J. Copley/Visuals Unlimited, (cr bl)Harry Taylor/DK Images, (bc)Mark A. Schneider/Photo Researchers, (br)Mark A. Schneider/Visuals Unlimited; **11** (inset)Patricia K. Armstrong/Visuals Unlimited, (r)Dennis Flaherty Photography/Photo Researchers; **13** KS Studios; **14** (l)Mark Burnett/Photo Researchers, (c)Dan Suzio/Photo Researchers, (r)Breck P. Kent/Earth Scenes; **15** (inset)Icon Images, (t)Bud Roberts/Visuals Unlimited, (b)Charles D. Winters/Photo Researchers; **16** (l)Andrew McClenaghan/Science Photo Library/Photo Researchers, (r)Charles D. Winters/Photo Researchers; **17** (t)Goeff Butler, (bl)Doug Martin, (br)Photo Researchers; **18** Matt Meadows; **19** Reuters NewMedia, Inc./CORBIS; **20** (Beryl, Spinel)Biophoto Associates/Photo Researchers, (Emerald, Topaz)H. Stern/Photo Researchers, (Ruby Spinel, Tanzanite)A.J. Copley/Visuals Unlimited, (Zoisite)Visuals Unlimited, (uncut Topaz)Mark A. Schneider/Visuals Unlimited; **21** (Olivine)University of Houston, (Peridot)Charles D. Winters/Photo Researchers, (Garnet)Arthur R. Hill/Visuals Unlimited, (Almandine)David Lees/CORBIS, (Quartz, Corundum)Doug Martin, (Amethyst)A.J. Copley/Visuals Unlimited, (Blue Sapphire)Vaughan Fleming/Science Photo Library/Photo Researchers; **22** (l)Francis G. Mayer/CORBIS, (r)National Museum of Natural History/©Smithsonian Institution; **23** (inset)Doug Martin, (l)Fred Whitehead/Earth Scenes; **24** (t)Matt Meadows, (bl)Paul Silverman/Fundamental Photographs, (br)Biophoto Associates/Photo Researchers; **25** Jim Cummins/Getty Images; **26** Matt Meadows; **27** (inset)José Manuel Sanchis Calvete/CORBIS, (t)Doug Martin, (bl)Andrew J. Martinez/Photo Researchers, (br)Charles D. Winter/Photo Researchers; **28** (bkgd)Science Photo Library/Custom Medical Stock Photo, (bl)Bettmann/CORBIS; **29** José Manuel Sanchis Calvete/CORBIS; **30** R. Weller/Cochise College; **32** José Manuel Sanchis Calvete/CORBIS; **33** Breck P. Kent/Earth Scenes; **34–35** Michael T. Sedam/CORBIS; **36** (l)CORBIS, (r)Doug Martin; **37** (tl)Steve Hoffman, (cl)Brent Turner/BLT Productions, (r)Breck P. Kent/Earth Scenes; **38** (bkgd)CORBIS/PictureQuest, (t)CORBIS, (bl)Martin Miller, (bc)Jeff Gnass, (br)Doug Sokell/Tom Stack & Assoc.; **39** Russ Clark; **40** USGS/HVO; **41** (t)Breck P. Kent/Earth Scenes, (b)Doug Martin; **42** (basalt)Mark Steinmetz, (scoria, obsidian)Doug Martin, (pumice)Tim Courlas, (others)Breck P. Kent/Earth Scenes; **44** (l)Breck P. Kent/Earth Scenes, (r)Doug Martin/Photo Researchers; **45** (t)Breck P. Kent/Earth Scenes, (l)Breck P. Kent/Earth Scenes, (bl)Courtesy Kent Ratajeski & Dr. Allen Glazner, University of North Carolina, (br)Alfred Pasieka/Photo Researchers; **47** (l)Aaron Haupt, (r)Robert Estall/CORBIS; **48** Paul Rocheleau/Index Stock; **49** (l)Timothy Fuller, (r)Steve McCutcheon/Visuals Unlimited; **51** (l)Icon Images, (cl)Doug Martin, (cr)Andrew Martinez/Photo Researchers, (r)John R. Foster/Photo Researchers; **52** (l)Breck P. Kent/Earth Scenes, (r)Aaron Haupt; **53** (bkgd)Georg Gerster/Photo Researchers, Icon Images; **55** Beth Davidow/Visuals Unlimited; **56** (l)Icon Images, (r)Breck P. Kent/Earth Scenes; **57** (l)Jack Sekowski, (r)Tim Courlas; **58** (bkgd)Y. Kawasaki/Photonica, (inset)Matt Turner/Liaison Agency; **60** Breck P. Kent/Earth Scenes; **61** Jeremy Woodhouse/DRK Photo; **64–65** Bill Ross/CORBIS; **67** Visuals Unlimited; **70** (l)George Lepp/CORBIS, (r)Carson Baldwin Jr./Earth Scenes; **71** Paul A. Souders/CORBIS; **72** (bkgd)Ian R. MacDonald/Texas A&M University, (l)Emory Kristof, (r)National Energy Technology Laboratory; **73** Hal Beral/Visuals Unlimited; **75** Roger Ressmeyer/CORBIS; **76** Spencer Grant/PhotoEdit, Inc.; **77** Inga Spence/Visuals Unlimited; **78** Robert Cameron/Stone/Getty Images; **79** Vince Streano/CORBIS; **80** (t)David Young-Wolff/PhotoEdit, Inc., (b)Earl Young/Archive Photos; **81** Peter Holden/Visuals Unlimited; **83** Aaron Haupt; **84** Joseph Nettis/Photo Researchers; **85** (t)Mark Joseph/Stone/Getty Images, (bl)Aaron Haupt, (br)Wyoming Mining Association; **88** (t)Aaron Haupt, (b)Joel W. Rogers/CORBIS; **89** Aaron Haupt; **90** (t)Ed Clark, (bl)Brown Brothers, (br)Shell Oil Co.; **91** (l)Andrew J. Martinez/Photo Researchers, (r)Coco McCoy/Rainbow; **95** Mark Joseph/Stone/Getty Images; **96–97** Bourseiller/Durieux/Photo Researchers; **100** Martin Land/Science Source/Photo Researchers; **103** Ralph White/CORBIS; **109** Davis Meltzer; **110** Craig Aurness/CORBIS; **112** Craig Brown/Index Stock; **113** Ric Ergenbright/CORBIS; **114** Roger Ressmeyer/CORBIS; **116** AP/Wide World Photos; **118** L. Lauber/Earth Scenes; **124–125** Chuck Nacke/TimeLife Pictures/Getty Images; **126** Tom & Therisa Stack; **128** (t)Tom Bean/DRK Photo, (b)Lysbeth Corsi/Visuals Unlimited; **129** David Parker/Photo Researchers; **130** Tom & Therisa Stack; **132** Robert W. Tope/Natural Science Illustrations;

Credits

139 (l)Steven D. Starr/Stock Boston, (r)Berkeley Seismological Laboratory; **140** AP/Wide World Photos; **141** David J. Cross/Peter Arnold, Inc.; **144** James L. Stanfield/National Geographic Image Collection; **145** David Young-Wolff/PhotoEdit, Inc.; **147** Reuters NewMedia Inc./CORBIS; **148** (tr)Richard Cummins/CORBIS, (l)Bettmann/CORBIS, (br)RO-MA Stock/Index Stock; **149** (l)Science VU/Visuals Unlimited, (r)Peter Menzel/Stock Boston; **152** Vince Streano/CORBIS; **154–155** Art Wolfe/Getty Images; **155** KS Studios; **156** Sigurjon Sindrason; **157** (t)John Cancalosi/DRK Photo, (b)Deborah Brosnan, Sustainable Ecosystems Institute; **161** Image courtesy NASA/GSFC/JPL, MISR Team; **162 163** Gary Rosenquist; **164** (bkgd)API/Explorer/Photo Researchers, (t)Krafft/HOA-QUI/Photo Researchers, (bl)Robert Hessler/Planet Earth Pictures, (br)Paul Chesley; **165** (l)Steve Kaufman/DRK Photo, (r)Dee Breger/Photo Researchers; **167** (t)Krafft/Explorer/Science Source/Photo Researchers, (b)Darrell Gulin/DRK Photo; **171** (tl)Joyce Photo/Photo Researchers, (tr)Doug Martin, (b)Brent Turner; **172** (t)Dick Canby/DRK Photo, (b)Tom Bean/DRK Photo; **174** Larry Ulrich/DRK Photo; **175** Amanita Pictures; **176** (t)Spencer Grant/PhotoEdit, Inc., (b)Darrell Gulin/DRK Photo; **177** Mimmo Jodice/CORBIS; **178** (l)Soames Summerhays/Photo Researchers, (r)Photri/The Stock Market/CORBIS; **179** Krafft/Explorer/Science Source/Photo Researchers; **181** Kerrick James/Getty Images; **184** PhotoDisc; **186** Tom Pantages; **190** Michell D. Bridwell/PhotoEdit, Inc.; **181** (t)Mark Burnett, (b)Dominic Oldershaw; **192** StudiOhio; **193** Timothy Fuller; **194** Aaron Haupt; **196** KS Studios; **197** Matt Meadows; **198** (t)Matt Meadows, (b)Doug Martin; **199** Doug Martin; **201** Amanita Pictures; **202** Bob Daemmrich; **204** Davis Barber/PhotoEdit, Inc.